Robert Drumm

Plants, Trees, Seeds

Suited for Southern Cultivation

Robert Drumm

Plants, Trees, Seeds
Suited for Southern Cultivation

ISBN/EAN: 9783741136276

Manufactured in Europe, USA, Canada, Australia, Japa

Cover: Foto ©berggeist007 / pixelio.de

Manufactured and distributed by brebook publishing software
(www.brebook.com)

Robert Drumm

Plants, Trees, Seeds

1889

ROSES, PLANTS,
TREES, SHRUBS,
SEEDS, ETC.

PLANTS
TREES
SEEDS

SUITED FOR SOUTHERN CULTIVATION

OFFERED BY

Robert Drumm & Co.,

Successors to DRUMM & BAKER BROS.,

FORT WORTH and DALLAS, TEXAS.

OFFICES:
800 Main Street, Fort Worth.
828 Main Street, Dallas.

NURSERY:
One and One-half Miles
East of Fort Worth.

Dissolution.

The co-partnership heretofore existing between Robert Drumm, Wm. Baker and J. B. Baker under the name and firm style of Drumm & Baker Bros., has this day been dissolved by mutual consent, ROBERT DRUMM & Co. succeeding to the business at the same stand. The latter will assume all debts of the late firm, and collect all claims due the firm.

DRUMM & BAKER BROS.

FORT WORTH, TEXAS, *Jan. 10, 1889.*

A Card.

Referring to the above notice, we relinquish our interest and good will in the late firm of Drumm & Baker Bros., to ROBERT DRUMM & Co., and bespeak for our successors the same liberal share of patronage heretofore extended the old firm.

WM. BAKER,
J. B. BAKER.

Notice.

Referring to the above notices, we would take this method of notifying our friends that we have purchased the business of the late firm of Drumm & Baker Bros., and will continue the business at the old stand. We hereby tender our sincere thanks to our former patrons, asking old friends to continue their patronage, and new acquaintances to favor us with their orders, all of which shall have the same prompt and careful attention as in the past, together with increased efforts to please and satisfy our customers.

ROBERT DRUMM & CO.

Introductory.

WE TAKE PLEASURE in presenting to our many friends our new General Catalogue, again enlarged and improved. We were highly gratified at our success last season, and take this opportunity to thank all our friends for their liberal support, kind words and letters. Almost without exception the expression is, in substance: "We have at last found a place in Texas, where for the same money, we receive larger and better plants, delivered in less time and in better condition, than from any northern firm." To meet the rapidly increasing trade, we have built more greenhouses in Fort Worth, and opened a branch house in Dallas. Last year we boasted of having the largest greenhouse establishment in Texas. We are striving now to so improve it that it shall be second to none in the country.

For our peculiar Texas climate we doubt if a better list of plants can be found anywhere than the one in this catalogue. Every season we try the novelties, discarding the inferior ones and retaining and offering only the best. Many desirable varieties have been introduced during the last few years. This is especially the case with roses, geraniums, chrysanthemums, plums and grapes. That we keep pace with the times is shown in the fact that at the Texas State Fair at Dallas, October, 1888, we received sixteen premiums, including the first Sweepstakes for Best Collection of Plants and first Sweepstakes for Best Collection of Cut-Flowers. Above were awarded over thirteen other exhibitors.

PLANTS BY EXPRESS.

We do not ship plants by mail in Texas, where we have such excellent facilities for shipping by express. By express we can ship much larger plants, without shaking all the dirt from the roots, so that when they arrive at their destination, they are as fresh and healthy as when first taken from our greenhouses. Still it takes considerable advice and reasoning to induce customers to order their plants shipped by express when they themselves have to pay the express charges, whereas if the plants come by mail the florist must prepay the postage.

To overcome this difficulty, namely to sell large, healthy plants, delivered at your homes, for the same price at which you can get the small "mailing" size,

We will deliver Free of Express Charges, to any Railroad Town in Texas, all orders for Plants amounting to $5 and up. Cash in all cases must accompany the order, to receive this advantage.

We except all plants *in pots*. Of course this offer does not include Flower Pots, Hanging Baskets, Vases, etc.

REASONS WHY YOU SHOULD BUY FROM US.

First. A glance at our Catalogue will show that we have as good an assortment as most of the leading florists in the United States.

Second. These plants are mostly varieties adapted to this climate.

Third. They have a more natural and healthy growth, not being forced by a continuous fire heat during the winter, as is the case in the north.

Fourth. Our prices are as low as those of any other first-class firm, and

Fifth. You receive them *free of express charges*, and only a few hours after they leave the greenhouses.

TERMS, CASH WITH ORDER.

Explicit Directions should be given for marking and shipping, giving the name in full, town, county and state. If the nearest express or freight office is different from the post office, it should be stated.

How to Send Money. Remittances should be made by Money Order or Registered Letter on Fort Worth, or draft on Fort Worth or New York. If private checks are sent, add 25 cents for collection.

Orders Should be Plainly Written on a separate sheet (use our blank order sheet), and not included in the body of the letter.

To parties who may visit our city, we extend a cordial invitation to call at our place of business on Main Street, or at the nursery in "Sylvania," and examine our stock.

CLUB OFFERS.

Combine Your Orders. Although our prices are extremely low, considering the quality of plants furnished, yet we offer the following additional inducements to persons who will exert themselves in our behalf by soliciting their neighbors to unite with them in ordering plants of us, and thus increase the amount remitted.

Packing Club Orders. Each person's order will be selected as desired and tied separate, so there is nothing to do but deliver the bundle.

Club Orders are shipped in one lot to the originator of the club.

For	$10.00	plants to the	value of	$12.00	may be	selected.
"	15.00	"	"	18.00	"	"
"	20.00	"	"	25.00	"	"

Individual Orders for above amounts are entitled to these discounts. Although we send some extras of our own selection with every order, we offer no premium on orders amounting to less than ten dollars.

Address all communications to

ROBERT DRUMM & CO.
Successors to Drumm & Baker Bros.
Fort Worth, Texas.

BRANCH,
Dallas, Texas.

Greenhouse Department.

NEW ROSES OF SPECIAL MERIT.

These are all valuable, and tested sufficiently to establish their claims fully.

METEOR.

The darkest red Hybrid Tea rose yet introduced; the color is a dark deep velvety crimson (no purple tinge to it, as is the general rule among its class); a constant bloomer, retaining its color a long time. 20 cents.

MAD. SCHWALLER.

A great bloomer, producing quantities of large, fine pink blooms; shape like a Hybrid. 20 cents.

MRS. JOHN LAING.

A fine rose, especially recommended for bearing; very fine shape and delicious fragrance; color a beautiful soft pink; a great bloomer. 30 cents.

QUEEN OF QUEENS.

With blush edge, large and of a perfect form; very full. A true Perpetual; every shoot brings a rose. Splendid for the garden. 35 cents.

PAPA GONTIER.

Extra large, finely formed buds, and flowers full and fragrant and of a beautiful brilliant carmine, changing to pale rose; reverse of petals fine purplish red. 15 cents.

PURITAN.

A splendid white rose "producing hybrid flowers on an everblooming bush." 25 cents.

SUSANNE BLANCHET.

An elegant bedding variety. Flesh salmon, shaded with salmon rose and white. A grand rose. 20 c.

SOUV. VICTOR HUGO.

Bright China rose, with copper yellow center, ends of petals suffused with carmine. A charmingly beautiful combination of coloring. 20 cents.

SOUV. GABRIELLE DREVET.

Salmon pink, with center of coppery rose; of good size and fine form. Very satisfactory in habit, growth and freedom of bloom. 20 c.

SOUV. THERESE LEVET.

(New Red Tea Rose.)

Color clear scarlet, shading darker; flowers remain in halfopened condition for a long time; will prove valuable for summer flowers. 15 c.

VISCOUNTESS FOLKESTONE.

A beautiful pink Hybrid Tea rose. 25 cents.

Puritan.

Princess Beatrice.

PRINCESS BEATRICE.

A new and beautiful rose; very free flowering; pale yellow, center golden yellow, petals edged with pink. 25 cents.

PRIMROSE DAME.

A fine rose of good form and habit; color, apricot crimson and white, forming a beautiful contrast. 20 cents.

ROSES

THE ROSES OFFERED in this catalogue are healthy young growing plants in two and three-inch pots. Most of them will commence blooming as soon as planted, and continue in bloom the whole season. All classes of the rose grow to perfection in Texas, and anyone, by spending a dollar or two, can have an abundance of bloom from February to December.

Roses delight in rich soil, plenty of water, and good cultivation. They should be pruned severely every season to produce the finest flowers.

We can supply large plants from the open ground of nearly all the varieties named below, until March 15th. As we grow the small roses in pots, they can be shipped and transplanted at any time. We would advise our friends, however, to plant early, so that the roses, can become thoroughly established in the ground before the dry weather sets in.

Plants from two-inch pots, except where noted, 10 cents each, $1 per dozen; three-inch pots, 20 cents each, $3 per dozen; large plants from open ground, 35 cents each.

EVERBLOOMING ROSES.

UNDER THIS HEADING we class the Tea, China and Bourbon Roses. They will all bloom as long as they grow; so to have flowers continually they should be kept in a growing condition all spring, summer and fall. In this way beautiful roses can be gathered from February to December.

Adam. Good; carmine pink; large and double; delicious tea scent.

Archduchess Isabella. White, shaded rose; a good bedder and constant bloomer. 15 cents.

Agrippina. Rich velvety crimson, beautiful in bud. For bedding it is unsurpassed.

Archduke Charles. Brilliant crimson scarlet, shaded with violet crimson; splendid.

Appoline. Clear pink; extra fine rose; grows with remarkable vigor, and is very hardy.

America. A large, fine flower, best in bud; dark, creamy yellow, changing to coppery or orange yellow; strong grower; suitable for trellis or pillar.

Aurora. Beautiful rosy blush.

Bourbon Queen. A splendid rose; large, fine form; very double, full and sweet; bright carmine, changing to clear rose, petals edged with pure white.

Bon Silene. Noted for the size and beauty of its buds, which are a rich deep rose; not double when full blown, but highly esteemed for its rich color and beautifully formed buds; very sweet.

Bella. Pure white; pretty pointed buds.

Crimson Bedder. Rich dark velvety crimson.

Cornelia Cook. Beautiful creamy white; buds of immense size and very double; sometimes does not open well, which is its weak point.

Comtesse Riza du Parc. A finely formed, highly colored flower; coppery rose, heavily shaded with carmine; fragrant and vigorous.

Charles Rovolli. Pure rose color; soft and pleasing. 15 cents.

Mrs. John Laing. (*See page 3.*)

Catherine Mermet.

Catherine Mermet. A beautiful variety, of bright flesh-colored rose; flowers large, full and globular, very double and sweet. A first-class rose in every respect.
Coquette de Lyon. Sometimes called "Yellow Hermosa." Pale yellow, free bloomer.
Cramoisi Superieur. Medium size; rich dark velvety crimson; very double, full and beautiful; compact growth; one of the best for bedding.
Douglass. Dark cherry red, rich and velvety; large, full and fragrant; desirable.
Doctor Berthet. Beautiful silvery pink, brightening at center to deep carmine, passing to rosy crimson; fine large flowers, very double and sweet.
Devoniensis. (Magnolia Rose.) Beautiful creamy white, with rosy center; large, very full and double, delightfully sweet Tea scent; one of the finest Roses. 20 cents.
Duchess de Brabant. Soft rosy pink; petals edged with silver.
Etoile de Lyon. Beautiful chrome yellow, deepening at the center to pure golden; flowers large, double and full. 15 cents.
Gloire de Rosamond. Rich velvety crimson, flamed with scarlet; full medium size.
Hermosa. Light pink; large, full and double; grows freely and blooms very profusely.
Homer. Rose, center salmon.
Isabella Sprunt. Bright canary yellow; large, beautiful buds; very sweet tea scented.
Louis Phillippe. Dark velvety crimson.
La Pactole. Creamy yellow center; very free flowering; a beautiful rose.
La Princess Vera. Flowers large and full, perfectly double; color creamy white.
Mad. Lambard. Salmon pink, shaded deep rose.
Mad. Bravy. Rich creamy white, with blush center; of perfect form.

Mad. Camille. Of large size, double and full; immense buds of delicate rosy flesh, changing to salmon rose, elegantly shaded and suffused with deep carmine; fragrant.
Mad. Welche. Beautiful amber yellow, deepening towards the center to orange or coppery yellow; extra, large, fine globular form, very double and full. 20 cents.
Mad. C. Kuster. White, orange yellow center.
Mad. Joseph Schwartz. White, beautifully flushed with pink and cream; of medium size, cupped and borne in clusters; one of the hardiest for out-door bedding. 15 cts.
Mad. Falcot. Apricot yellow; flower of medium size and fullness, but beautiful in bud.
Mad. Russell. Creamy white, shaded with pink.
Mad'lle Rachel. A lovely tea rose; pure snow white, very double and deliciously scented; makes beautiful buds; is an elegant rose for either house or open ground.
Marie Ducher. Rich transparent salmon, with fawn center; large, double and sweet.
Mad. Brest. A dark rosy red; double and large.
Mad. de St. Joseph. Fawn, shaded salmon; large, full, sweet-scented; fine.
Mad. Margottin. Deep citron yellow, center rosy peach; large, very full and globular.
Marie Guillot. Perfection in form; the flower is large and double to the center; grows well and is healthy; pure white when fully open.
Marie Van Houtte. Of a lovely pale yellow color, with the outer petals suffused bright pink, and the inner ones edged rose; large, full, fine of form. 20 cents.
Niphetos. Pure white, fine buds; a poor grower out-doors, but good for pot culture.
Perle des Jardins. The finest and most popular dwarf yellow rose in existence. Flowers canary yellow, often golden yellow; of the most beautiful form, and large.
Pink Daily. Clear bright pink; medium size; free and double; fragrant.
Queen of Bedders. Very rich, dark velvety crimson; very free blooming, flower compact and full; one of the best crimson bedders; a grand sort. 20 cents.
Regulus. Large perfect form, full and double; carmine, with purple and rose shading.
Safrano. Bright apricot yellow, changing to orange and fawn, sometimes tinted with rose; highly valued for its beautiful buds; exceedingly profuse in bloom.

Souv. de la Malmaison. The flower is extremely large, quartered and double to the center; flesh white. 15c.
Sombreuil. This magnificent variety has immense finely formed flowers of a beautiful white, tinged with delicate rose.
Souvenir d'un Ami. Fine delicate rose shaded with salmon; very large, full and double; exquisitely fragrant.
Sunset. A sport from *Perle des Jardins*, which it closely resembles except in color; the color is a remarkable shade of rich golden amber, elegantly tinged and shaded with dark ruddy copper; intensely beautiful; with us has not proved as vigorous a grower as its parent. 25 c.
The Bride. The new white Tea rose that has become so popular. We consider it among the few new roses worth planting in the South. It is of unusually fine form, being a sport from *Catherine Mermet*. 15 cents.
Vallee de Chamounix. Coppery yellow, tinged with rosy blush. 20 cents.
William Francis Bennett. A new rose; valuable for winter forcing, but almost useless for out-door culture. Dark crimson; large and beautiful in bud. 15 cents.
White Tea. Same as *White Daily*; pure white; a constant and profuse bloomer; fragrant and desirable.

HYBRID PERPETUAL ROSES.

THIS CLASS is not so popular in the South as the Everblooming kinds, yet some of them excel all other classes in size of flower and richness of color. They bloom profusely in their season, and some varieties produce also a few fine flowers later. The shades of crimson in this class are especially fine; but so far no varieties of a yellow tint have been produced in Hybrid Perpetuals.

Anna de Diesbach. Clear rose; fine color; large and showy.

Baroness Rothschild. The form of this rose is absolute perfection; it is globular in shape, the petals curved and of waxen texture; flowers large, with satiny finish and of a very pleasing shade of delicate but decided pink. 25 cents each; large plants, 75 cents.

Belle de Normandy. Beautiful clear rose, shaded and clouded with rosy carmine and lilac; very large and sweet.

Captain Christy. Delicate flesh color, shaded rose to center; a large, finely formed flower.

Caroline Goodrich. Rich velvety crimson.

Gabriel Tournier. A fine free blooming rose; light crimson, shaded deep rose.

Earl of Pembroke. A strong grower; very soft, velvety crimson, enlivened on the edge of petals with bright red. 20 cents.

General Jacqueminot. Bright crimson, exceedingly rich and velvety; the buds and partially expanded flowers are very fine, and fairly glow with deep rich coloring.

General Washington. Rosy carmine, shaded scarlet; very brilliant and glossy; double and of good form.

Giant of Battles. Brilliant crimson; large, full and well formed.

John Hopper. Fine bright rose; flowers very large, cupped, full and well formed.

La Reine. Beautiful clear bright rose; very large.

Mad. Plantier. Pure white flowers in clusters; summer bloomer.

Mabel Morrison. Of excellent habit and a very free bloomer; petals thick and of a soft, smooth texture, shell-shaped; very double and rounded in form; of the purest white, and almost velvety in its finish. One bloom, with its foliage, is a bouquet in itself. Large plants, 75 cents.

Mad. Alfred de Rougemont. Pure white, with light rose shading; flowers not very large, but full.

Mad. Chas. Wood. Rosy carmine, sometimes darker; large and double to the center; of open form.

Magna Charta. Pink, suffused with carmine; full and globular. One of the very finest.

Mad. Masson. Reddish crimson; large and double; a good bloomer. 15 cents.

Mad. Rivers. Clear pink; very full and sweet.

Mervielle de Lyon. White, sometimes slightly tinted flesh; very large and double. One of the best white roses grown, and always a favorite in its beauty. Large plants only, 75 cents.

Prince Camille de Rohan. Dark crimson maroon, very rich; moderately double; of good habit and fine in bud, but a shy bloomer.

Paul Neyron. Pink; very large; one of the best of this class, and probably the largest rose grown.

Prince of Wales. Rich crimson; a vigorous and good bloomer, of rich coloring. 20 cents.

Pius IX. Clear bright rose, changing to rosy pink, delicately shaded; very large, fragrant and desirable; has been popular for many years.

Merveille de Lyon.

HYBRID TEA ROSES.

THIS CLASS of roses is the result of a cross between the varieties of Hybrid Perpetuals and Teas. The great advantage claimed for them is that they combine the free flowering qualities of the Tea class with the rich coloring and to some extent the hardiness of the Hybrid Perpetuals. Some varieties of great value are in the section.

American Beauty. (Or *Red La France.*) Resembles *La France* in shape, size and freedom of bloom, but is of a deep shade of rose. The growth is also different, being quite robust, and the flowers are garnished by leaves close to them. 20 cents.

American Beauty.

Beauty of Stapleford. Flowers large, double, and handsomely formed; bright pink, shading gradually toward the center to deep rosy carmine. 15 cents.

La France. Well known; perhaps the best rose in existence. "For the few who do not already know it, we will say it is of extra size, extra double and of superb form. No rose can surpass it in delicate coloring—silvery rose, shaded with pink: in fragrance it is incomparable: in form, perfect. The sweetest and most useful of all roses."

Antoine Verdier. Bright glowing pink, beautifully shaded with rich crimson; large and globular, very full and sweet; a constant and profuse bloomer.

Duke of Connaught. Buds extra large, very full and fragrant; long and finely formed; brilliant scarlet.

Pearl. Rosy flesh, of medium size, full; has pretty buds; an elegant sort. 15 cents.

Triomphe d'Angers. Bright, fiery red, changing to darkest velvety crimson, tinged with purple. 15 cents.

POLYANTHA ROSES.

Anna Marie de Montravel. A beautiful pure white rose; very double, perfect flowers; very sweet-scented; a real little gem. 15 cents.

Little Pet. Pure white flowers; always in bloom; very pretty. 15 cents.

M'lle Cecile Brunner. Quite distinct from the others, and very pretty, with slender salmon colored buds, heavily shaded with rose; the buds are the true tea rose form, distinct from others of this class.

Mignonette. Its color is clear delicate rose, with a frequent and peculiar intermingling of white that makes it unique and pleasing; blooms in large clusters, and is very beautiful.

Perle d'Or. A beautiful yellow variety of this much prized class of roses. Like the rest of the class, it is of dwarf habit, and very free flowering. It blooms in clusters, often numbering from thirty to forty flowers each. Color of buds beautiful nankeen yellow, with vivid orange center, each petal tipped white, changing to rose. 20 cents.

Polyantha Rose.

CLIMBING TEAS AND NOISETTES.

Chromatella or Cloth of Gold. Golden yellow; fragrant, large and beautiful; 20 cts.

Gloire de Dijon. One of the finest roses grown. It is noted for the great size of its flowers, its delicate tea scent, and its exquisite shades of color, being a blending of amber, carmine and cream. 15 cts.

Lamarque. Flowers of medium size, borne in large clusters—established plants bear thousands of blossoms; pure white, double.

Marechal Neil. A beautiful deep sulphur yellow, very full, large and exceedingly sweet. It is the finest yellow rose in existence. It has a climbing habit, and when allowed to grow until it attains a large size, as it will in a few years, it yields thousands of beautiful flowers. Large budded plants, $1 each; smaller plants, 10 to 25 cts.

Mad. Alfred Carriere. Flesh white, with salmon yellow at the base of petals.

Reine Marie Henriette. Bright cherry red, of a pure shade; a strong, vigorous grower.

Climbing Hermosa. Same as *Hermosa*, but of climbing habit.

Reine Olga de Wurtemberg. Large, full and fine; rosy flesh, delicately tinged with salmon yellow, petals margined with crimson; blooms in clusters.

Marechal Neil.

William Allen Richardson. Orange yellow, outer petals lighter; center copper yellow; very rich.

Washington. Of medium size; pure white, very double; blooms profusely in large clusters; is a strong grower, suitable for trellises, etc.; quite hardy.

Woodland Margaret. White; vigorous.

HOW OUR PLANTS PLEASE.

ABILENE, TEXAS, *April 27, 1888.*

The flowers came in fine condition. I am delighted with them. The extras are lovely.

MRS. T. L. ODEM.

EL PASO, TEXAS, *April 16, 1888.*

My last order arrived safely, and gave full satisfaction, the plants growing right straight along. Many thanks for the gifts that you added.

MRS. ELIZABETH HEIDERHOFF.

CHAPPELL HILL, WASHINGTON, CO., TEXAS.

I am so delighted with my plants that I now send you another order. I am so sorry that I did not know of your firm sooner, for the plants that I have received from the north are so slow getting started off. Many thanks for the extras. One of them is blooming beautifully.

MRS. FANNIE A. LIDE.

Type of the Newer Double Geraniums.

GERANIUMS.

NO CLASS of plants has yet been found to take the place of the Geranium. There has been so much improvement in them in the last few years, that they now can be had in all the shades of scarlet, crimson, pink, salmon, white and striped. They are indispensable in any collection of plants, either for pot culture or for bedding out. For bedding out we recommend the solid darker shades, especially the single scarlets. They glory in the hottest sun, and will stand considerable drouth. They stand the hot sun of the South better than any other class of plants. They produce more flowers and make a better display in whatever place they are grown, than anything that could be grown in a similar place. All who saw our brilliant beds of Geraniums last summer, were surprised to learn they had never been watered.

PRICE OF GERANIUMS.

Except where noted, named varieties, vigorous, healthy plants, 10 cents each, $1 per doz.; assorted, unnamed varieties, 75 cts. per doz., $5 per 100; plants from three-inch pots, 15 cents each, $1.50 per dozen; large blooming plants in five to seven-inch pots, 25 to 50 cents each.

DOUBLE FLOWERING GERANIUMS.

Asa Gray. Light salmon pink; dwarf; very free flowering and an excellent bedder. $6.50 per 100.
Alba Perfecta. White, tinged pink.
Amazon. White.
Apple Blossom. Rosy salmon pink.
Banquise. Rose and pure white.
Belle France Rich purple, bordered amaranth, upper petals spotted white.
Bonant. Brilliant sparkling vermilion.
Coquet. White.
Chas. Darwin. Rich violet purple.
C. H. Wagner. Crimson.
De Brazza. Bright madder orange. 15 c.
Double Red. This is not a distinct variety but a collection of the best double scarlet and crimson varieties. A bed of Double Reds on our grounds, produced more fine blooms than any other on the place. As we grow great numbers of them we sell them at 75 cents per dozen, $5 per 100.
Delobel. Bright scarlet.
Effective. Brightest scarlet, base of petals white.
Ernest Lauth. Deep violet; large truss.
Etoile de Roses. Beautiful bright China rose.
Grand Chancellor Faidherbe. Dark soft red, shaded maroon.
Gambetta. Dark red.
Gen. Billot. Rosy scarlet.
Gen. J. A. Garfield. Soft rose.
Gen. Millot. Dark soft red; extra fine. 15 cents.
Gil Blas. Currant red, striped fiery red.
Hofgartner Eichler. Large truss of salmon, shaded bronze.
Henry Cannell. Intense scarlet.
Hazel Kirke. Crimson purple.
Iroquois. Fiery scarlet.
J. H. Klippart. Scarlet.
J. Y. Murkland. Rose center, margined white.
Le Notre. Dark rosy violet.
L' Abbe Gregoire. Amaranthine red.
L' Elysee. Rose color.
Lemoine's Cannell. Deep purple, suffused crimson and scarlet.
Lena Woods. Crimson scarlet.
La Niagara. Fine pure white.
Lolita Pena. Beautiful lively magenta.
L' Andalouise. Pure white.
La Fraicheur. Tender lilac rose; a new shade, and quite distinct. 15 cents.
La Victorie. Pure white; full and fine.
Mon. J. Chretien. Fiery red velvet color.
M. l' Abbe Jalabert. Lively amaranth with fiery spots.
M. A. Piola. Currant red, mixed with carmine and orange. 15 cents.
Mad. Ed. Andre. Salmon with bronze.
Mrs. Hayes. Rosy pink.
Mrs. Cope. Rich carmine red.
Mrs. E. G. Hill. Pale blush, shaded lavender.
Maonissa. A lovely shade of rose; strong grower and good bloomer.
Mrs. E. J. Bane. Rich crimson.
M. Dubus. Light rose.
M. Hardy. Large flower of lilac and tender rose.
M. Puteaux Chaimbault. Fine rose.
Mrs. John Thorpe. Scarlet, shaded maroon.
Mad. Thibaut. A beautiful rich rose color, shaded with carmine violet; one of the best double bedders of its color.
Peter Henderson. Orange scarlet.
Orange Perfection. Enormous trusses of orange vermilion flowers; a constant bloomer in-doors or out. 15 cents.
Ruby Triumph. Crimson scarlet.
Remarkable. Bright crimson, very free; good truss.
Richard Brett. Peculiar light scarlet.
Robt. George. Deep crimson scarlet; large.
Stanislaus Malingre. White and pink.
The Blonde. Bright salmon and pink.

SINGLE FLOWERING GERANIUMS.

Anna Scott. Deep crimson, shaded maroon.
Aurora Borealis. Orange flame color.
Bamford's Glory. Bright clear scarlet.
Bishop Simpson. A very large variety, with large trusses of rich salmon flowers.
Clement Boutard. Flesh colored.
Celestial. Rose, tinged with white.
Cyclope. Large trusses; white, shaded salmon, orange center.
Cosmos. Salmon and orange. 15 cents.
Genome. Soft bright scarlet.
Gen. Grant. Dazzling scarlet, large truss, one of the best. $6 per 100.
Gen. Sherman. Bright scarlet.
Kate Patterson. Beautiful clear salmon orange.
Miss Blanche. Deep purplish pink.
Mrs. J. P. Anthony. White, with rosy salmon.
Mrs. Garfield. White.
M. Chevreul. Bright amaranthine red.
Mary Hallock Foote. Clear pale salmon, with large white eye.
New Life. Bright scarlet, striped with white.
Neve. White.
Pres. Simon. Bright clear red.
Pres. Frazier. Rose and pink.
Pabelon. Scarlet.
Queen of the West. Bright orange scarlet; has large trusses and is a profuse bloomer. We know of no finer Geranium for planting out in beds.
Streak of Luck. A rival of New Life. Salmon, streaked with white; dwarf, but a most profuse bloomer.
Scarlet Bedder. This, like our "Double Red," is not a distinct variety, but a collection of the best single scarlets, selected from the best and healthiest beds, regardless of names. They are all well grown healthy plants. 75 cents per doz., $5 per 100.

GERANIUMS, NEW VARIETIES.

We take great pleasure in offering the following collection of Geraniums, knowing that for profusion of bloom and size of flowers they are unsurpassed. Some of the newer varieties are immense. The following are selected from the latest European importations, and the cream of recent American novelties.

15 cents each, $1.50 per dozen, except where noted.

New Double Geranium, White Swan.

NEWEST DOUBLE GERANIUMS.

Baron de St. Dedier. Orange, bordered white; large truss of finest form.

Ct. Turati. Large truss; very large round florets; brilliant flame color. Desirable.

Contre-Amiral Knoor. Cherry, shaded with lilac, upper petals distinctly marked with orange. A first class sort.

Contraste. Flowers and trusses very large; lower petals orange and carmine, upper and center petals clear orange, making a contrast of colors that is very beautiful. 20 cents.

Edward Koch. Flowers very large, brilliant cinnabar color at the center. Very beautiful and one that is valuable

Esperance. Flowers full and of the same beautiful color as rose "Malmaison," with a salmon center. A beautiful variety.

Fratelli Ferrario. Bright orange apricot color. A variety that we can recommend highly.

Gen. de Courcy. Enormous sized spherical trusses, flowers very full; brick red in the center, changing to a salmon rose, marked with white. 20 cents.

Gloire de France. Large round florets, of waxy appearance; colors carmine and white, with carmine center; each petal delicately edged with carmine. An excellent pot variety, to which its habit is finely suited—a fine sort in every way. 25 cents.

Jules Lartigue. Flowers semi-double; upper petals carmine, lower violet lake; plant a vigorous grower.

Heros d' Usagara. Orange cinnamon; beautiful form.

Les Huguenots. Trusses large; florets very double, silvery lilac.

Mad. Guillot. Color a very distinct and beautiful China rose. 20 cents.

Mad. Guilbert. Very large florets, borne in immense sized trusses of a very beautiful pure rose color. An extra fine Geranium.

Marvel. Dark crimson maroon.

Mad. Hoste. Color flesh, shaded white and salmon. One of the finest geraniums in cultivation, and of very rich effect. 25 cents.

Progression. A strong growing variety, bearing trusses of well formed flowers, of orange scarlet, on the style of *Jealousy*; very distinct.

Palmyre. Immense sized trusses of well formed white flowers; a fine bedder and a good geranium.

Princess Dutailly. Very dark amaranth, bordered with maroon, upper petals marked with purple.

Princess d' Anhalt. Pure white; truss large and firm.

Refinement. White, delicately edged with rose color.

Sergeant Bobillot. Large truss; beautiful rose veined violet.

Ville de Lausanne. Currant, lightly shaded.

Volcan. Fiery red, trusses and flowers large.

White Swan. A perfectly double pure white variety; a remarkably free bloomer.

NEWEST SINGLE GERANIUMS.

Arc-en-Ciel. Reddish lake, tinted fiery red at the border of the petals, the upper petals distinctly marked with orange, 20 cents.

Etincelant. Flowers large; very brilliant and pure capucine red; trusses of great size. An exceedingly fine Geranium.

Francois Arago. Flowers finely formed; a free bloomer, on strong foot-stalks; silvery salmon, shaded peach. 20 cents.

Favorite. Large fine florets of extra fine form; trusses of immense size, beautiful cherry scarlet; plant of good habit and very floriferous.

Frau Louise Voith. The grandest single pink Geranium out. Color beautiful clear and brilliant carmine rose with a very large and pure white eye; florets round and of perfect form; trusses of enormous size and beautiful shape. 25 cts.

Gloire de Lorraine. Trusses very large; colors rich orange shaded with salmon, and distinctly marked with white.

Himalaya. Flowers large; orange salmon bordered with clear salmon with a large clear white eye; plant dwarf and very free.

Henri Martin. Clear orange scarlet; the largest truss of this color. An excellent bedding sort.

Jumbo. Immense truss of velvety crimson. One of the finest.

Jules Ferry. Brilliant dark red, enormous truss.

New Louis Ulbach. Brilliant red; very bright and rich.

Poete Nationale. Delicate pink, deepening to soft peach bloom.

Newland's Mary. A better variety in every respect than the old *Master Christine*, which has hitherto brought forth more expressions of satisfaction than any other zonale. 20 cents.

M. Albert Delaux. Color beautiful flesh distinctly and beautifully specked or sanded with carmine rose, upper petals marked white.

Renan Truss large; flowers medium size, salmon apricot color.

Reflector. Very bright scarlet, with large pure white eye; trusses large and freely produced.

Rembrandt. Upper petals rich velvety crimson, lower petals being deeply shaded with rich violet.

Starlight. Pure white, with broad pink center; distinct pure white eye.

Sam Sloan. An extra fine velvety crimson, carrying immense trusses in great quantities; an excellent bedder, unrivaled for producing masses of deep bright red bloom.

Swanley Gem. Color rosy salmon red, with large white eye.

VARIEGATED-LEAVED GERANIUMS.

Their rich striking foliage makes them desirable for baskets, stands, vases or bordering plants, and for massing many of them are unsurpassed. Our list includes the best varieties grown in the two and three color sections, both bronze and silver leaved.

15 cents each, $1.50 per dozen, except where noted.

Corinne. Clear golden yellow foliage, flowers bright double scarlet.

Crystal Palace Gem. The reverse of *Happy Thought;* center of leaf green, with broad band of light yellow. One of the best of its class.

Chieftain. Fine showy foliage having a golden disc surrounded by a wide bronze zone.

Distinction. Foliage green with a narrow jet-black zone; very neat and distinct.

Freak of Nature. An improvement on *Happy Thought.* Novel, distinct and attractive. 20 cents.

Golden Chain. Center of leaf green, with maroon border of yellow.

Happy Thought. Center of leaf creamy yellow, with a broad margin of deep green; flowers magenta; dwarf habit.

King of Bronzes. Leaves light yellow with very broad chocolate zone.

Mountain of Snow. Green, edged with pure white.

Mrs. Pollock. Leaves green, crimson, scarlet and yellow; a most beautiful sort for house culture.

Mad. Salleroi. The leaves are deep olive green in the center, with a broad white margin; compact habit, symmetrical form, dense foliage and dwarf habit. Good for pots or borders. 10 cents each, $1 per dozen, $6 per 100.

Marechal McMahon. Light yellow, with bronze zone.

Silver Cord. An improvement on the old *Mountain of Snow.*

Zulu. Bright yellow with deep chocolate zone. 20 cents.

Freak of Nature.

Ivy-Leaved Geranium.

IVY-LEAVED GERANIUMS.

This class of Geraniums, from their graceful growth and beauty of flowers and foliage, are favorites wherever grown. They are remarkably adapted for vases or baskets, rock work, etc., and also desirable for pot culture in the house. The past few years has brought us many fine sorts.

10 cents each.

Anna Pfitzer. Flowers very-large, semi-double; clear rose.

Comte Horace de Choiseul. Plant vigorous, leaves without zones; flowers very large, double and imbricated, in good trusses; bright golden salmon.

Comtesse Horace de Choiseul. Beautiful satiny rose, shaded towards edge of petals white; most charming variety.

La Rosiere. Flowers of perfect shape, double and of good size; satin rose; an excellent variety.

L'Elegante. An extremely pretty variety of bright green foliage, with broad band of creamy white, often margined with pink; dense clusters of pure white flowers. This sort makes a splendid house plant, and is very fine in every way. 20 cents.

Spalding's Pet. An excellent old variety, of erect habit and compact form, profuse in bloom; soft cerise red.

Lucy Lemoine. Flowers nearly white, veined with purple.

SWEET-SCENTED GERANIUMS.

10 cents each, $1 per dozen, except where noted.

Apple. Round light green leaves, delightfully fragrant; a universal favorite. 25 c.

Attar of Roses. Fine cut-leaved; deliciously rose scented.

Balm. Very large foliage; fragrant.

Dr. Livingston. Finely cut foliage, rose-scented.

Lemon. Small, crisp leaves; strongly lemon scented.

Nutmeg. Free flowering; small leaves resembling those of the apple-scented.

Oak-Leaved. Large leaves, with a broad black mark. 15 cents.

Peppermint. Large leaves, finely scented. 15 cents.

Rose-Scented. Fine cut leaves.

Rose-Scented. Large broad leaves. Both varieties of the Rose Geranium make excellent bedding plants, and are always admired. $1 per dozen.

Mrs. Taylor. Scarlet flowered rose; a distinct variety of the scented Geranium, with a strong rose fragrance, and large, deep scarlet flowers. Combining, as it does, free-flowering qualities with fragrant foliage, it is very useful for summer cut flowers, and as a pot plant for winter, cannot be excelled. 15 cents.

Shrubland Pet. Small miniature growth, rosy red flowers.

Transit. Sweet rose scent; leaves remain fresh a long time. 15 cents.

PELARGONIUMS OR LADY WASHINGTON GERANIUMS.

A beautiful class of plants for spring and early summer blooming. We offer several varieties of this class.

15 cents each, $1.50 per dozen.

☞ We grow and can furnish a number of other good varieties, but have not the space to describe them here.

General Collection of Plants.

ABUTILONS.

Showy plants, and easily grown in almost any soil; very free flowering and pretty.

10 cents each, except where noted.

A. Besson. Flowers long and pendulous on long stems; light orange red, with lighter veinings.
Diadem. Wine color, veined violet.
Etendard. Deep red, shaded crimson.
Firefly. Brilliant red, shaded crimson.
Lustrous. Brilliant dark red; free bloomer. 15 cts.
Laura Powell. Bright golden yellow.
M. Delaux. Light red, center white; large flowers.
Purpurea. Beautiful shade of purple; a new shade.
Purity. Pure white. 15 c.
Pluton. Deep red, with dark veinings.
Rosalita. Flesh color, shaded rose.
Robert George. Deep orange, streaked crimson.
Thompsoni Plena. Double flowers; very deep orange, shaded and streaked with crimson; leaves mottled yellow and green. 15 cts.

Abutilon.

AGERATUM.

Very useful plants for bedding or borders, flowering continually during the summer. By cutting back and potting in the fall they will continue to flower all winter.

10 cents each, $1 per dozen.

Lady Jane. A distinct, compact growing variety; flowers deep blue.
Mexicanum Variegata. Leaves variegated with creamy white, flowers blue; desirable.
White Cap. Flowers pure white; strong compact grower.
Mayflower. A very dwarf variety, and the best for carpet bedding, only growing four inches high; dark blue flowers. 15 cts.

Anthemis Coronaria.

ANTHEMIS CORONARIA.

A pretty bedding plant, equally adapted to growing in pots; flowers golden yellow; resembles a double Feverfew. 15 cts. each.

ACHANIA MALVAVISCUS.

A profuse flowering plant, producing its scarlet flowers at the end of every branch. 10 cts.

ASCLEPIAS LINIFOLIA ALBA.

A herbaceous plant, bearing umbels of pure white flowers during the entire summer, when planted out. 10 cts.

ACHYRANTHUS.

Very pretty and attractive leaved plants; useful in massing and ribbon gardening, as they contrast well with Coleus and other foliage plants. They are easily raised, and stand our hottest sun, besides bearing any amount of neglect. In height from one foot to eighteen inches. Excellent for the center of vases and baskets, and very satisfactory for window culture, when kept in bushy form by pinching back.

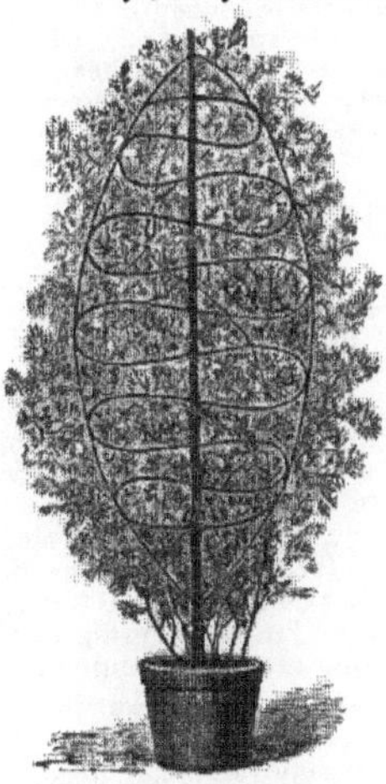
Asparagus Tenuissimus.

10 cents each, $1 per dozen.

Lindenli. Leaves of a deep blood red color, narrow, elliptical, reflecting varying tints of red and purple.

Formosum. Rich bright yellow, with light green edge and crimson mid-rib, and crimson stems; very fine.

Aureus Reticulatus. Green, netted with yellow, sometimes dashed with crimson; pink stems and veins; fine.

Gilsonii. Leaves carmine, with the stems of a deep shade of pink; dwarf and dense growth.

ALOES.

A valuable class of plants for rock work in summer, or the house in winter. 25 cts.

ASPARAGUS GENUISSIMUS.

An elegant evergreen climber from South Africa, with slender, smooth stems and numerous spreading branches. A handsome ornamental plant; its plumy, feathery growths are very useful for decoration. 20 cts.

ALOYSIA CITRIODORA.

(Lemon Verbena.)

A favorite garden plant, with delightfully fragrant foliage; fine for bouquets and for spreading through linen presses. We never yet met a person who was not fond of its aromatic fragrance. It attains a large size in a single season; may be wintered in the cellar, not allowing it to become absolutely dry. Grows well in almost any soil. 15 cts.

ALTERNANTHERAS.

Beautiful dwarf plants of compact habit, growing about six inches high, and admirably adapted for edgings to flower beds or ribbon lines, their beautifully variegated foliage of crimson, purple, yellow, pink and green, forming rich masses of color, always attractive and highly ornamental. We grow thousands of Alternantheras, because they do so well in this climate, and are certain to give satisfaction wherever planted. A bed of Alternantheras forms a perpetual carpet of bright colors.

10 cents each, 75 cents per dozen, $4.50 per hundred.

Aurea Nana. Bright golden yellow, holding its color well the entire season.

Amœna Spectabilis. Crimson, pink and brown; the finest.

Versicolor. Foliage ovate; olive, crimson and chocolate; when used for edgings they must be frequently sheared; the young tips will then assume the most brilliant hues.

AMARYLLIS.

Splendid flowering bulbs, suitable only for house or greenhouse culture.

Belladonna. Red, fine showy flowers. 35 cts. each.

Purpurea. Brilliant purplish scarlet. 60 cts. each.

Formosissima. (Scarlet Jacobean Lily.) Summer flowering; rich crimson velvet. 25 cts. each.

Sweet Alyssum. A constant bloomer, either in the house or open ground. 10 cts. each, 75 cts. per dozen.

ASTERS.

Plants grown from choicest seed. 75 cts. per dozen.

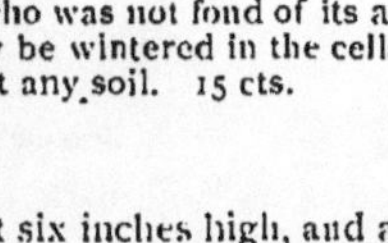
Am. Formosissima.

ASTILBE JAPONICA.

One of the finest hardy garden plants; when in bloom it is about one foot in height. The flowers are produced in branching, feather-like spikes of pure white. A fine cemetery plant, and beautiful for any place. It is largely forced in in winter by florists for its elegant, plumy white flower-spikes. 25 cts.

Begonia Rubra.

BEGONIAS.

This beautiful class of of plants is deservedly popular. The beauty of their foliage, combined with their graceful flowers and free blooming qualities, tend to make them a most desirable class of plants. Grown as pot plants for summer or winter decoration, they have but few equals; they are also very useful for baskets or vases, or for bedding out in half shady places on the lawn, flowering profusely during the whole summer.

15 cts. each, except where noted.

Alba Perfecta Grandiflora. This variety closely resembles Begonia Rubra in foliage and growth; has beautiful pure white blossoms.

Bruantii. Foliage a very bright green; quite large; pure white flowers, borne in great profusion. 25c.

Digswelliana. Flowers dark crimson, center pink; very free bloomer; a beautiful variety.

Dreggii. Pure white flowers, freely produced; fine foliage; a very pretty sort.

Hybrida Multiflora. Rosy pink.

Metallica. A fine, erect-growing Beg mia, with dark, rough leaves; the surface is a lustrous bronze-green; veins depressed and dark red. A free bloomer The panicles of unopened buds are bright red, with surface like plush; when open the flower is waxy white. This Begonia is a splendid house plant.

McBethii. Of the Weltoniensis type, with very deeply indented leaves, which are fine and small. Shrubby in growth, and very free flowering, being constant the year round. White flowers, waxy in texture, and carried in panicles.

Ricinifolia. Very large palmate leaves, and showy, rose purple flowers; stems red; leaves silvery and dark green marked. 25 cts.

Sanguinea. The deep red-leaved variety; one of the handsomest grown. 20 cts.

Sutton's White Perfection. This is a seedling of the Semperflorens class. In this Begonia we have the finest white both for market and cut-flower purposes we have seen. It is a continuous bloomer both summer and winter, and is beautiful in cut-flower work. Plant of strong robust growth; foliage dark green; flowers large and of a beautiful pearly white, borne on long stems. 20 cts.

Robusta. Bright carmine flowers; very free bloomer.

Rubra. This is one of the finest acquisitions to our winter flowering plants; the leaves are of the d rkest green, the flowers large, ruby red, glossy and wax-like. This peculiarity is so marked that when plants are placed singly in a room the glossy appearance of the leaves and flowers gives the idea that they are artificial rather than natural. Is popular wherever known, and will please all who grow it. 15 cts.

Semperflorens Alba. An old and popular variety, with bright glossy green foliage and white flowers. Stands the sun well, and is always in bloom. 10 cts.

Sandersonii. Scarlet; fine for winter blooming.

Schmidtii. A new variety with bronze-colored leaves; of dwarf, dense growth, with a profusion of rosy white flowers.

Weltoniensis. Exceedingly fine, shrubby habit, with strong green leaves and bright crimson stems; flowers waxy pink, very profuse in winter and summer; a distinct and desirable sort; stands the sun well; a good bedding sort.

BEGONIA REX.

Grown for the great beauty of their foliage. The leaves are large, beautifully variegated, and marked with a peculiar silvery metallic gloss. 25 cents.

Group of Cacti.

BOUVARDIAS.

These are among the most important plants for winter blooming. For bedding they are excellent, beginning to bloom in July and continuing until frost, when they can be lifted and put in the house, where they will continue to flower through winter. 20 cts.

CACTUS.

Cacti are prized, not only for their curious appearance, but also for their blooming qualities, many of them producing flowers of remarkable beauty. Keep them dry and warm and they will thrive. We have a good collection of the best kinds from Southern and Western Texas. 15 to 50 cents each; assorted varieties, $3 per dozen.

CALADIUMS.

Esculentum. (Elephant's Ear.) The most striking distinct ornamental foliage plant in cultivation; desirable for pot or tub culture, and fine for bedding out. With a plentiful supply of water, the leaves may be grown from four to six feet long. 25 cents.

FANCY-LEAVED CALADIUMS.

For ornamental foliage plants for summer, nothing can equal the gorgeous coloring of the Fancy Caladiums. They are of easy culture, thriving best in pots or boxes in the house or in shaded places out-of-doors.

25 cents each, except where noted.

Annibal. Crimson veins on green ground; spotted with carmine.
Berose. Deep green, spotted red, veined crimson; light green center.
Boildieu. Bright crimson center, deep green margin.
Candida. Clear white ground, strongly marked ribs.
Chantini Splendens. Spotted rose and crimson.
Clio. Ground color deep rose, shaded white; green ribs and margin. 40 cents.

Fancy Caladium.

Diana. Rich green spotted, red center, crimson rayed.
E. G. Henderson. Green, spotted rose, rayed crimson.
Gerard Dow. Pale yellow, red veins, carmine mid rib.
La Perle du Bresil. White and green; transparent.
L'Albane. White, with green veins; large.
Leplay. Green, marked white, veined crimson.
Mad. F. Kochelein. White ground, violet ribs, green veins.
Marjolin Shaffer. White, veined rosy lake. 40 c.
Mon. J. Linden. White with metallic reflections, coral-rose veins and green border. 40 cents.
Mad. A. Bleu. Green, with white blotches and scarlet veins.
Meyerbeer. White, veins green, mid-rib red.
Napoleon III. Flamed crimson center on green ground.
Sanchoniatum. Crimson center and ribs, pea-green margin.
Sieboldii. Fiery red center, spotted red.
Velasquez. Carmine ground, crimson veins, pink spots, green border. 40 cents.
Ville de Mulhouse. Green and white, shaded rose; rich green center.

CAMPSIDIUM FILICIFOLIUM.

An elegant hard-wooded vine, with foliage as beautifully divided as any fern frond. Few vines are so beautiful or so excellently adapted to house culture. Its hard-wooded character enables it to withstand the atmosphere of the room where other plants would suffer, and being nearly hardy, it is not injured by a low temperature. 20 cents each.

Canna.

CEREUS GRANDIFLORUS.

(Night-Blooming Cereus.

Flowers straw color, remarkable for their beauty and sweetness; they begin to open about 7 P. M., and continue until after midnight, and are from nine to eleven inches across. 15 cents.

CENTAUREA GYMNOCARPA.

(Dusty Miller.)

Silvery gray foliage, contrasting well with dark foliage plants, in ribbon lines; also form a very pretty specimen plant in a pot, or in a vase or rustic baskets, and very good. 10 cents each, 75 per dozen.

CANNAS.

Large, showy plants, which grow about four feet high, with broad, richly-colored leaves, giving them a great value in semi-tropical gardening. They look best in groups, and can be handsomely combined with Caladiums and similar plants. The flowers are borne in spikes at the top of the plant, and are usually of the shades of orange, red and yellow. 15 c.

CALLA ETHIOPICA.

The well known Egyptian Lily, or Lily of the Nile. A fine house plant of the easiest culture; requires plenty of water during the growing season, from October to May. Dry bulbs, 25 cents each; large plants in pots, 50 cents to $1 each.

CALLA RICHARDIA MACULATA.

A plant belonging to the same order as the Calla Ethiopica, with beautifully spotted leaves. It flowers abundantly during the summer months, planted out in the open border. The flowers are shaped like those of a Calla, and are pure white, shaded with violet inside. It is a deciduous plant; kept dry in winter and starts in spring like a Dahlia. 20 cents.

CESTRUM PARQUI.

(Night Blooming Jessamine.)

A plant of strong shrubby growth, with small greenish-white blossoms, with a delightful odor, which is dispensed freely during the night only; of easy cultivation. 15 to 25 cents each.

CLERODENDRON.

Balfouri. A beautiful hot-house climber; flowers borne in pendant clusters; corolla dark scarlet, the tube encased in a pure white sack-like calyx. 25 cents.

Fragrans. Plant of dwarf habit; flowers pinkish white, double and very fragrant. 25 c.

CARNATIONS.

Carnations are quite hardy in this latitude, and may be left out all winter. Planted out in April they will commence flowering in early summer and will continue until checked by heavy frosts in winter.

Except where noted, 10 cents each, $1 per dozen.

Buttercup. Deep rich golden yellow, with a few clear streaks of carmine. 20 cents.

Crimson King. Dark crimson-maroon.

Chas. Sumner. Large bright pink, fringed.

Dawn. Delicate rosy salmon, changing to pure white at the edge. 20 cents.

Dora Copperfield. Dark maroon with scarlet shadings; very large double flowers. 25 cents.

Hinze's White. One of the finest white Carnations in cultivation; flowers very large and of perfect shape; color white, with a delicate tint of pink.

Field of Gold. Pure yellow, with no stripe., 20 cents.

Grace Wilder. Delicate pink; free bloomer. 15 cents.

John McCullough. Fine scarlet; extra.

La Purite. Color beautiful pink; free bloomer.

E. G. Hill. The finest scarlet yet introduced, of vigorous, healthy growth and very free. 20 cents.

Mrs. W. P. Brady. Yellow, striped crimson.

Pres. De Graw. Pure white; vigorous and free.

Pres. Garfield. Rich vermilion; good bloomer; flowers large and of good form.

Sec. Windom. Intense vermilion scarlet.

The Century. Strong healthy grower and constant bloomer; rich glowing carmine color, and strong clove scent. 15 cents.

Venus. Light canary yellow; robust and free. 15 cents.

CUPHEA PLATYCENTRA.

(Ladies' Cigar Plant.)

A choice well known everblooming plant, growing about twelve inches in height. A fine pot plant, useful for bedding in front lines in flower borders. Is of neat habit, loaded the entire season with scarlet, tubular, pendulous flowers. Worthy of a place in every flower garden. 10 cents.

CYPERUS ALTERNIFOLIUS.

A grass-like plant, throwing up stems to the height of about two feet, surmounted at the top by a cluster or whorl of leaves, diverging horizontally, giving the plant a very curious appearance. A splendid plant for the center of baskets, vases or Wardian cases, or as a water-plant. 25 and 50 cents.

CLEMATIS.

This family of plants is noted for their rapid but slender growth, delicate foliage and profusion of lovely bloom throughout the summer.

Coccinea. Perfectly hardy, and a rapid grower. The leaves are deep shining green; flowers bell-shaped, and of the most intense coral scarlet. 50 cents.

Jackmanni. This is perhaps the most popular of the new fine perpetual Clematis. The flowers are large, of intense violet purple, and remarkable for their intense velvety richness; a strong grower and hardy. $1 each.

Flammula. An old variety, highly prized for the fragrance of its small white flowers, and its small dark green leaves, which remain on the plant very late. 50 cents.

Crispa. Flowers produced singly on long stalks, abundant, and one and a-half inches long; lilac purple and delightfully fragrant. 50 cents.

Lady Caroline Neville. Fine flowers from six to seven inches in diameter; delicate blush white, with a broad purplish lilac band in the center of each sepal. 75 cents.

Miss Bateman. A magnificent plant, both in growth and flower; the blooms are large, of good shape; pure white, banded with creamy white down the center of each sepal. 75 cents.

Henryi. A magnificent large white flower; a free grower and most profuse bloomer. 50 cents.

Lanuginosa. Light azure blue. 50 cents.

Jean d'Arc. Flowers white, large and perfect, with three pale blue stripes on each petal. 60 cents.

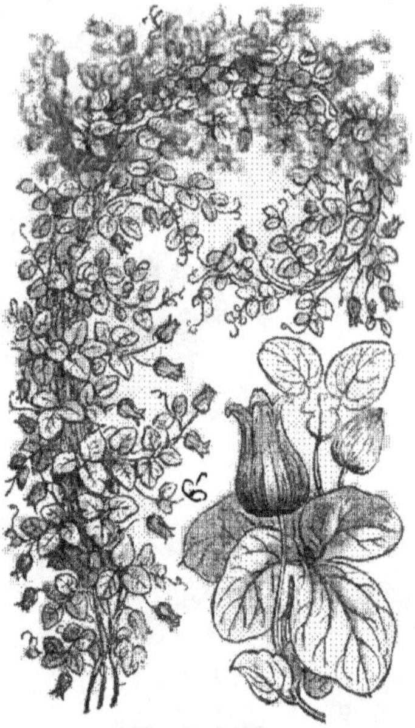

Clematis Coccinea.

COLEUS.

10 cents each, 75 cents per dozen.

Asa Gray. Beautifully mottled green, crimson and white.

Captivation. The deeply toothed leaves are bright green, with a feathered center of sulphur yellow, which is flushed with a pale tint of rosy purple.

Flambeau. Richly colored, exceedingly attractive; the leaves have a broad surface of rich velvety maroon, with bar of bright magenta, bordered by a narrow edge of olive.

Firebrand. Maroon, flamed and shaded with brilliant fiery red.

Golden Bedder. Bright golden yellow.

John Goode. Yellow.

Clare. Velvety crimson, green margined.

Kentish Fire. Bright carmine; crimson and green serrated edge.

Coleus.

Midnight. Maroon, flamed crimson.

Mrs. Cooper. Golden yellow, green margined.

Mrs. J. Schultz. Bright golden yellow ground, with bright scarlet and carmine markings.

Mrs. Humphreys. Light claret, rimmed yellow.

Negro. Very dark foliage.

Prince of Prussia. Crimson scarlet, yellow margin.

Progress. Dark olive green, blotched with purple, crimson and gold; the most distinct yet introduced.

Rainbow. Maroon, yellow edge.

Speciosa. Green, with light yellow center.

Setting Sun. Rich bronze crimson, bright golden edge.

Spotted Gem. Yellow, blotched crimson, green and orange; very effective.

Tesselata. Evenly margined with green and yellow; strong grower.

Verschaffelti. Rich velvety crimson; one of the best for bedding out.

Zebra. Yellow, crimson and green streaked; striking.

Comte de Germny.

CHRYSANTHEMUMS.

WITH THE exception of the rose, no flower has gained so extensive a popularity as the Chrysanthemum. Never in the history of any flower has so much interest been manifested as there is in this flower of to-day. The Chrysanthemum is the people's flower, as it appeals directly to the heart, and bears but slightly on the purse.

Our list is the largest and most complete to be found anywhere in Texas. It contains none but good varieties, and embraces the best of the old varieties that were popular many years ago, together with the best and newest varieties of the past two years.

Always plant Chrysanthemums in an open spot where they will have sunshine each day. Make the soil rich with manure or bone dust; liquid manure may be given occasionally. Stake each plant, tying it loosely, so that the wind will not injure the branches. Pot-grown plants delight in great quantities of water. If for bloom indoors they should be lifted about the 1st of October and potted. Set in a shady place for a few days after thoroughly drenching with water. Afterwards expose them to the full light, but do not keep warmer than fifty degrees.

10 cents each, $1 per dozen, except where noted; one hundred, in a hundred different varieties, $6.50.

JAPANESE VARIETIES.

The following list has no "chaff" in it, but includes the very best varieties, both of American and European origin. We shall be glad to make selections for our friends who are unacquainted with the sorts.

The Japanese varieties are, of course, the most variable in form and coloring, and are perhaps the most admired. The list of varieties we offer has been selected as the most distinct.

Abd-el-Kader. Rich deep maroon crimson; petals twisted; a beautiful and distinct variety; large and fine.

Belle Paule. Very large; center of each petal purest white, distinctly edged with rose.

Blanche Neige. One of the largest and purest white flowers; a magnificent flower—a gem.

Bras Rouge. Velvety crimson-maroon, with reverse of petals deep bronze.

Bicolor. Enormous large flat flowers; red, striped with orange. 15 cents.

Christmas Eve. Magnificent white; each petal curves and twists, the whole forming a ball of peculiar appearance; the latest of all Chrysanthemums—good at Christmas.

Comte de Germiny. Very large flowers and remarkably broad petals of a rich orange brown; reverse of petals silvery bronze. The parent of many of the finest sorts grown.

David Allen. Very large; chrome-yellow outside; center crimson red. 15 cents.

Duchess. Enormous red flowers; very free and distinct.

Delicatum. Blush; very large; petals broad and flat, tapering to a point.

Domination. A grand variety, with erect petals; blush, with rosy base, slightly fringed.

Dormillion. Very large pæony-shaped flowers of a beautiful amaranth color, reverse of petals white.

Eucharis. Outer petals broad and reflexed, of the purest white; the center deep yellow, with a corona-like circle formed above the general surface, which is delicate creamy white; distinct and fine; late. 15 cents each.

Elaine. Pure white; one of the very best.

Gloriosum. Very light lemon, with immense flowers, having narrow petals most gracefully curved and twisted; very early. 20 cents.

Frizon. Pure golden yellow; flowers large and perfect; a very early variety.

Fimbriatum. Most delicate pink, fringed; a general favorite. 20 cents.

Fernand Feral. Soft rose, shaded mauve and suffused cream color; large and fine.

Fabias de Maderanaz. Large center; blush and bronze, with long twisted petals of creamy white; beautiful anemone-formed.

Fair Maid of Guernsey. Flowers large, of snowiest white, in clusters; one of the best white varieties.

Glorie Rayonnante. Quilled florets of a clear satiny rose color with lilac shade; flowers very large; early.

Grandiflorum. Flowers of immense size, often six inches in diameter; petals very broad, incurving, so as to form a solid ball of the purest golden yellow. 20 cents.

Gorgeous. Golden yellow; a magnificent variety; early and distinct.

Grace Floyd. Deep rose color, large and free.

Hackney Holmes. A gem among the very best; ground color deep bright crimson, with tips of pure golden yellow.

Hon. John Welsh. Dark lake; a new color in Chrysanthemums; free and good.

H. Waterer. Enormous flower of great substance, flat; yellow, with copper center. 20 c.

J. Collins. Immense large flat flowers of copper bronze; a self-colored variety.

Jessica. White, very long petals; shows a lemon eye when fully expanded; a very good bloomer.

Juvena. Flowers of the richest, deepest crimson; petals narrow and much twisted; center of each is golden yellow.

Julius Cæsar. A very distinct color of an entirely new shade, being red orange, or the very lightest chestnut; the flowers are large, rather smooth in outline. 15 cents.

Jennie Y. Markland. Very large, having a flat surface; rich golden yellow, shaded apricot and rose. 15 cents.

Kira Kana. A rich tone of pure chrome, of the largest size, in large heads.

Kata Kanka. Very large; rich, deep bronzy buff, with a peculiar warm chrome shade.

Lord Byron. A magnificent large variety; very distinct orange, tipped with red. 15 cts.

Lady Selborne. A very large pure white variety of the greatest merit; quite early.

La Favorite d' Exposition. White, tinged with pink; petals long and twisted.

Mrs. Geo. W. Childs. Of elegant incurved shape; outside petals are white; dark rose inside.

M. Planchenau. Mauve, shaded rose and silver; early; extra.

Manhattan. Lavender rose, with a distinct white line down the center of each petal.

Moonlight. Immense flowers of pure white; very beautiful.

Mad. C. Audiguier. Flowers of the purest rosy pink; a gem.

Mrs. J. B. Wilson. This is probably the finest white Japanese Chrysanthemum ever sent out. Beautiful ivory white; of immense size, being fully seven inches in diameter, and of unusual substance; petals very broad and long, giving the flower a very bold and strong expression. 25 c.

Chrysanthemum, Mrs. J. B. Wilson.

Mrs. C. Carey. Pearly white, much curved and twisted.
M. Castel. Bright rich crimson, reverse of petals rich golden yellow.
Mrs. R. Brett. A distinct variety, differing from all other varieties in its peculiar plum-like flowers and rich coloring of pure gold; a gem.
Mrs. John Thorpe. Brilliant crimson, very decided in coloring; opens in whor's; a grand cut flower. 20 cents.
M. Cochet. Silvery white, suffused rose, reverse of petals carmine-violet.
M. Boyer. Petals very long and twisted; beautiful silvery lilac rose.
M. Buchard. Bronze; large and full.
M. Roux. Beautiful dark crimson, white center.
Pietro Diaz. Flowers of deepest red garnet with golden reflex. 15 cents.
President Garfield. Brightest carmine; large flowers and very distinct; a notable variety of great beauty.
President Cleveland. Delicate blush, changing to pure white; a fine variety. 15 cents.
Peter the Great. A most showy bright lemon-yellow, with beautiful foliage and habit.
Red Dragon. Bright red flowers, blotched and splashed with yellow; early.
Red Gauntlet. Rich crimson red; of good size, in compact heads.
Robert Bottomley. A magnificent white variety. Makes a fine specimen plant. 15 cts.
Source d' Or. Golden twisted florets, tipped yellowish brown; large flower.
Syringa. Lilac; of immense size, center petals increasing, other petals very irregular.
Sadie Martinot. Golden yellow.
Snowstorm. Pure white, distinct and free—hence its name.
Souv. de Mt. Blanc. Pure white, petals long and twisted; one of the very best. 15 cts.
Thunberg. Flowers very large; of a pure primrose shade of yellow.
Tokio. Deep reddish bronze, shaded yellow.
Talfourd Salter. Red and carmine.
Tubiflorum, Delicate pale rose mauve, passing to white; a flower of novel form and effect. 20 cents each.
White Dragon. Pure white.
W. K. Harris. This variety forms perfect balls of nankeen yellow; at first it shows a light red center. 20 cents.

CHINESE CHRYSANTHEMUMS.

In this group will be found the varieties that are smooth in outline and of regular shape. The Chinese section is growing rapidly into favor, and while perhaps there are not as many admirers of them as of the Japanese, they are fast getting deeper and deeper into the affections of the people.

Autumn Glow. Sulphur yellow; large and full.
Alfred Salter. Deep rosy pink; large and fine.
Barbara. Rich orange amber; large, and one of the most perfectly incurved.
Bouquet Blanche. Pure white; intermediate late.
Beverley. Creamy white; broad incurved florets of fine form.
Beauty of Stoke Newington. Beautifully incurved; lilac blush; fine for decoration.
Cullingfordii. Rich crimson, shaded scarlet; the flowers are very large and reflexed; fine and distinct. 20 cents.
Diana. Purest white; large and full; very fine.
Emily Dale. Pale straw color.
Fingal. Violet purple; broad petals, flowers quite globular; incurved, very fine.
Golden Empress. Primrose yellow; fine show flower; incurved.
Golden Beverley. Flowers large and perfectly incurved.
Golden Queen of England. Very large; rich lemon yellow; one of the best; incurved.
Guernsey Pride. Yellow; fine large blooms; incurved, extra.
George Glenny. Beautiful lemon white; very fine habit.
Golden Prince. Primrose yellow; very free.
Hero of Stoke Newington. Rosy blush, shaded purple.
Helvetic. Carmine, shaded amaranth, with silver center.

Incarnata. Clear rose, passing to white; very pretty and distinct; a good sort.
Jardin des Plantes. Bright golden yellow; splendid color; incurved.
Lady St. Clair. Pure white, large and full; early and fine.
Lord Wolseley. A grand variety; rich deep bronzy-red, shaded purple.
Lippincot. Large, pure white.
Margaret of York. Very fine sulphur yellow; extra.
Mrs. George Rundle. One of the most beautiful white Chrysanthemums in cultivation.
Mrs. N. Hallock. Snowy white; a really valuable acquisition.
Mr. Bunn. Probably this is the very finest of all incurved yellow Chrysanthemums.
M'd'lle Madeleine Tezier. White, delicately tinted blush.
Mad. Aristee. Beautiful quilled yellow flowers; a very distinct and pleasing variety.

Chinese Incurved Chrysanthemum.

New M. Roux. Pæony-shaped flowers of a pure chamois color; extra fine.
Verschagine. Early flowering; of a rich lilac rose, each petal pointed with white; flowers medium size, globular; very free flowering.
Webb's Queen. Soft silvery rose; early.

POMPON, ANEMONE AND QUILLED CHRYSANTHEMUMS.

Aequisition. Clear rose lilac; yellow center, very high and round; fine.
Alba Nana. Pure white; small, compact.
Anais. Rosy lilac, golden center.
Black Douglass. This is one of the handsomest Pompons. Large crimson flowers, very free; extra bright and cheerful. 25 cents.
Bob. Dark brown crimson, fine color; a great favorite, and fine for specimens.

Anemone Flowered.

Fanny. Maroon red, free and fine.
King of the Anemones. Rich crimson; petals.
L'Orangere. Beautiful clear apricot yellow, of an attractive shade; very free and fine.
Marabout. Beautifully fringed; white.
Model of Perfection. Rich lilac, edged pure white; very distinct and pretty.
M'lle Mathilde Reynaud. Rose, with yellow center, tipped with white.
Nellie Rainford. Buff.
Princess Louise. Delicate rosy lilac; full high center; fine; early.

SINGLE VARIETIES.

These are certainly the most valuable for cut flowers. They are little or no trouble to grow, produce many blooms, and last long.

Attraction. Long and pure white petals, shaded toward the edges with rose; center yellow. 15 c.
Blanche Coles. Pure white, with long slender petals. 15 cents.
Hamlet. Fine chestnut.
Mrs. Gubbins. Very large, creamy white, much twisted; fine habit; large yellow disc.
Mrs. Robertson. Very large, creamy white; petals much twisted.

DAHLIAS.

Well known autumn-flowering plants, growing from two to five feet high, and producing a profusion of flowers of the most perfect and beautiful form, varying in color from the purest white to the darkest maroon. Dry roots, 25 cents each, $2.50 per dozen.

Single Dahlia.

DAPHNE INDICA.

An evergreen greenhouse shrub, with fine pinkish white, deliciously sweet flowers; blooms from January to April. 50 cents.

DIANTHUS.

Double hardy pinks; of rich colors and sweet clover scent. 75 cents per dozen.

DIELYTRA SPECTABILIS.

(From China.)

One of the most popular of our tuberous-rooted plants. It is perfectly hardy, and equally adapted to out-door planting or forcing for early spring blooming. It will produce its showy racemes of delicate pink and white heart-shaped flowers from February to April in the greenhouse, flowering in the open ground in May and June. 25 to 50 cents each.

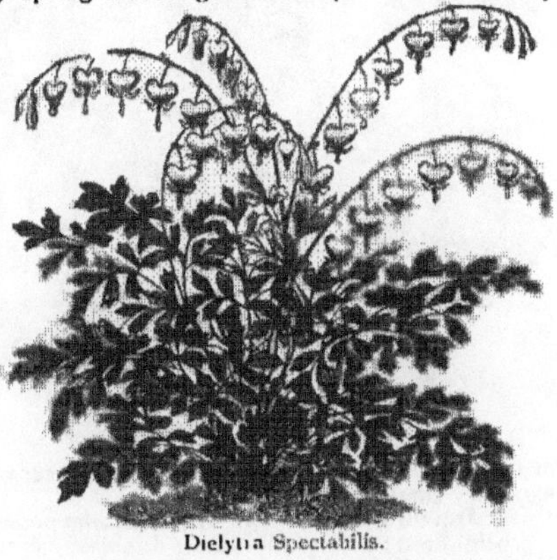
Dielytra Spectabilis.

DRACENA.

Beautiful ornamental-leaved plants, much used for centers of baskets or stands; they prefer shade.

Terminalis. Rich crimson foliage; the showiest of the Dracænas; very ornamental as a parlor plant or in rustic baskets, etc. 50 and 75 cents each.

Indivisa. Narrow grass-like foliage. 25 cents.

EUPHORBIA SPLENDENS.

This is a curious plant, having but few leaves, but is covered over with thorns one-half inch long; blooms freely in winter and summer; flowers scarlet, in clusters. 10 and 20 cents.

Ficus Elastica.

EUCHARIS AMAZONICA.

Amazonian Lily; large, pure white, fragrant flowers, full four inches in diameter, and are produced in abundance on strong plants. $1 each.

FICUS.

Elastica. (India Rubber Tree.) Very large, smooth, leathery leaves; evergreen foliage; generally esteemed one of the finest house plants grown, the plants attaining a large size and tree shape. A very fine plant for the lawn or bay window; not hardy. $1 each.

Australis. Smaller leaves than the above. $1 each.

FORGET-ME-NOT.

(Myosotis Palustris.)

It delights in partly shaded moist places; flowers in clusters, light blue. 10 cents.

FUCHSIAS.

Fuchsias delight in a rich, light soil, and may be grown either as pot plants, or planted out of doors. In all cases they must have plenty of water and be protected from the hot mid-day sun.

Fuchsia, Storm King.

10 cents each, $1 per dozen, except where noted.

Amiral Miot. Plant very bushy and free blooming; double corolla, clear prune color; sepals brilliant red. 15 cents.

Bulgaria. Sepals red; corolla violet prune; large, single.

Col. Borgnis Desbordes. Free blooming, dwarf; single corolla clear violet; sepals salmon rose.

Crepuscule. Tube and sepals dark red; corolla double, violet, striped with rose.

Drame. Corolla beautiful blue, large and double; sepals reflexed and red. 15 cents.

Duke of Albany. Corolla single, purplish red; tube and sepals red.

De Mirbel. Sepals bright red; large single corolla, rose and violet.

Enfant Prodigue. Flowers large; corolla double, violet blue; sepals short, red; plant very free. 35 cents.

Gustave Dore. Double white.

Joseph Rosain. Double purple.

Le Majestueux. Sepals white; corolla rosy carmine; flower single, plant vigorous. 25 cts.

Lustre Improved. An improvement on the old favorite; corolla brilliant scarlet; tube and sepals waxy white. 15 cents.

Mad. Thibaut. Double corolla, of bright carmine, bordered white; very pretty.

Model. Double white.

Paul Deroulede. Sepals bright scarlet; corolla double, large, violet, changing to bright red; a new color in the doubles. 35 cents.

Perle Von Brunn. Sepals clear red; corolla very double, of immense size and of the purest white. The finest double white yet offered. 20 cents.

Pres. F. Gunther. Large; double lilac.

Penelope. A grand single white.

Telegraph. Corolla violet; sepals red. 20 cents.

Speciosa. Orange carmine.

Storm King or Frau Emma Topfer. Sepals rosy coral; corolla clear blush; flowers double and large. 15 cents.

FEVERFEW.

Double White. A plant will please everyone; very free blooming, double daisy-like flowers; useful for summer bouquets. 10 cents.

Golden Feather. Golden colored foliage; very dwarf, fine for edging. 10 cents each, 75 cents per dozen.

FESTUCA GLAUCA.

A graceful, bluish green grass; fine for hanging baskets, etc. 10 cents.

FRAGARIA INDICA.

Sometimes called the Indica or False Strawberry. It belongs to the same class, and is like the strawberry in its habit—low and trailing—and the foliage handsome. Its fruit is scarlet and remains long upon the plant. Handsome for baskets or for windows. 15c.

FERNS.

Good assortment, 25 cents to $1 each. Handsome specimens of Maiden Hair Ferns, 50 cents to $1 each; smaller plants, 15 cents.

Collection of Ferns.

FARFUGIUM GRANDE.

A beautiful ornamental foliage plant; leaves nearlly round, of dark glossy green, with numerous cream-colored spots and blotches; flowers light purple. 30 cents.

GNAPHALIUM LANAGUM.

A downy white foliaged plant of creeping habit, admirably adapted for the front lines of ribbon borders; also a fine basket plant. 10 cents.

GLADIOLUS.

The Gladiolus is the most beautiful of the summer or tender bulbs, with tall spikes of flowers, some two feet or more in height, often several from the same bulb. The flowers are of almost every desirable color, brilliant scarlet, crimson, creamy white, striped, blotched and spotted in the most curious manner. Set the bulbs from six to nine inches apart, and about four inches deep. Plant from middle of March to first of June. It is a good way to plant at two or three different times, ten days or two weeks apart. This will give a succession of bloom from July to November. In the fall, before hard frost, take up the bulbs, remove the tops, leave to dry in the air for a few days, and store in some cool place, secure from the frost until spring. 10 cents each, 75 cts. per dozen.

HELIOTROPES.

These plants are universal favorites on account of their delightful fragrance. Flower equally well as bedding plants in summer, or as pot plants in the house during winter.

10 cents each, $1 per dozen, except where noted.

Evening Star. Light blue, very fragrant; a profuse bloomer, and one of the best for out-doors.

Mad. de Blonay. Large truss, nearly pure white. 15 cents.

Swanley Giant. Carmine rose; the size of truss is immense, often measuring ten inches in diameter; of the most exquisite fragrance. 15 cents.

Albert Deleaux. Foliage beautifully variegated green and yellow; flowers large, bright lavender. 15 cents.

Jersey Beauty. The best purple.

Pres. Garfield. Dark purple.

Mrs. David Wood. A new semi-double, with immense trusses, fine habit, free bloomer; flowers very lasting. 25 cents.

Violet Queen. Deepest violet purple, with long, almost pure white eye; very fragrant; vigorous habit and very floriferous; new.

White Lady. A strong growing, free branching plant; very profuse bloomer; flowers large and of the purest white.

Heliotropes.

HOYA CARNOSA.

(Wax Plant.)

A climbing plant, with thick, fleshy leaves, bearing flesh-colored, star-shaped flowers; one of the best plants for house culture, as it stands the extremes of heat and cold better than most plants, and is not easily injured by neglect. 25 cents.

HELIANTHUS MULTIFLORUS PLENUS.

A beautiful hardy plant, growing to the height of three to four feet; flowers a rich golden yellow, very double and as large as a medium-sized Dahlia. Much prized as corsage flowers. It begins to bloom in July and continues until frost. This should be in every collection of plants, as it certainly is one of the finest hardy yellow flowering plants in cultivation. 15 cents each, two for 25 cents.

Helianthus Multiflorus Plenus.

HIBISCUS.

A beautiful class of greenhouse shrubs, with handsome glossy foliage and large showy flowers, often measuring six inches in diameter; they succeed admirably bedded out during the summer.

15 cents each, except where noted.

Aurantiaca. Large double orange-colored flowers. 20 cents.
Collerii. Double flowers, buff yellow, with a scarlet base, very distinct; a new variety from the South Sea Islands. 20 cents.
Cooperii. Beautifully variegated foliage, white, green and pink.
Grandiflora. Rich, glossy foliage, with crimson scarlet flowers.
Rosa Sinensis. Bright red, single flowers.
Miniatus. Semi-double flowers, brilliant vermilion scarlet; very handsome.
Rubra. Double variety, with large red flowers.
Sub-Violaceous. This is the largest flowering of the Hibiscus family; of a beautiful shade of bright crimson, tinted with violet.
Versicolor. Very large single flowers, beautifully striped crimson, rose and white.
Schizopetalus. A beautiful and distinct kind, with drooping pendulous, reflected, orange red and lacinated petals; highly valuable for training to the pillars of greenhouses; of climbing habit. 25 cents.
Dennisonii. The most distinct of all Hibiscus. Color a delicate carmine, and as the flower gets older changes to pure white. 25 cents.

IVY.

English. The well known evergreen climber; quite hardy. 10 to 25 cents.
Senecio Scandens. (German or Parlor Ivy.) A more rapid growing and more succulent kind; well adapted for covering trellis work quickly or training in the parlor; leaves glossy green and flowers yellow, in clusters. 10 cents.
Kenilworth Ivy. (*Lunaria Cymbalaria.*) A neat and delicate plant of trailing habit, with small bright green ivy-shaped leaves, and small light violet-colored flowers; well adapted for hanging baskets, vases, etc. 10 cents.
Ground Ivy. A low creeping plant; suitable for rockeries, baskets, etc. 10 cents.

Moon Flower.

IPOMEA NOCTIPHITON.

(Evening Glory or Moon Flower.)

A rapid growing plant of the Morning Glory family, with pure white moon-like flowers, six inches in diameter, which open at night. As a rapid climber for covering arbors, verandas, trellises, trees or walls, it has no superior. 15 cts.

IMPATIENS SULTANI.

Of compact, neat habit, and a perpetual bloomer; the flowers are a peculiar brilliant rosy scarlet color, one and a-half inches in diameter, and produced very freely. 15 cents each, $1.50 per dozen,

IRIS.

Their rich colors, quaint forms and sweet perfume, render many of the Irises equal in interest and beauty to the Orchids. If they were better known they would be more extensively planted, as they are very cheap. When planted in clumps, and allowed to remain undisturbed, they improve in beauty every successive year. If you order any bulbs, try a few Iris, and you will be pleased with them.

Lantanas.

English. All colors. 50 cents per dozen.
Mont Blanc. The only pure white English Iris in cultivation. 15 cents each.
Spanish Mixed. 25 cents per dozen, $1.25 per 100.
Spanish Named. 10 cents each, 75 cents per dozen.
Pavonia. Pure white, blue spotted. 5 cents each, 50 cents per dozen.
German. 10 cents each, $1 per dozen.
Susiana. (The Large Mourning Iris.) Lilac, spotted black; very large flower. 25 cents each, $2.50 per dozen.
Tuberosum. (The Snake's Head Iris.) The interior petals black, edged with deep green. 10 cents each, $1 per dozen.

JASMINUMS.

Grandiflorum. From India. Flowers pure white, star-shaped, of exquisite fragrance; blooms from October until May without intermission. 15 cents.
Grand Duke. A variety with double, creamy white flowers; fragrant. 25 cents.

JUSTICIA.

A handsome fall and winter blooming plant, bearing large spikes of lilac and white flowers.

Carnea. Pink. 15 cents.
Hydrangoides. White. 15 cents.

LANTANAS.

This desirable class of plants are annually growing in favor, their brilliant colors, robust form and profuse blooming habit, render them worthy of a place in every flower garden. It is now becoming so generally cultivated that we rank it as one of our most important plants for bedding purposes during summer, thriving well in the hottest sun when many other plants suffer with the heat and drouth, affording a profusion of flowers embracing all the most delicate shades of orange, sulphur, creamy white, etc., from one color to another as they increase in age and development.

10 cents each, $1 per dozen, except where noted.

Aurantiaca. Large, orange red; fine.
Alba Perfecta. White; very fine.
Delicatissima, Lilac pink; trailing habit, neat growing and one of the very prettiest. 15 cents.
Golden Ball. Bright orange; one of the best bedders.
Harkett's Perfection. Foliage variegated, with yellow and lilac flowers.
Hendersoni. Flowers large, saffron changing to buff.

LOBELIA.

Among the most useful plants for hanging baskets, or for the front of outside row in ribbon lines. Their dwarf habit, and the profusion of their flowers, render them exceedingly ornamental.

Erinus Speciosa. Of trailing growth; flowers of superb ultra-marine blue. 10 cents.
Alba Maxima. Showy white flowers, of drooping and spreading habit. 10 cents.

LILIES.

Lilies do not receive the attention in Texas that they deserve. No flower is prettier or more easily grown. We offer the following varieties as being desirable for Texas.

Harrisii Lily.

Lilium Auratum. (Gold-Banded Lily.) The grandest of all lilies; has white flowers, spotted with maroon, and a golden band through the center of each petal. 25 to 50 cents.

Candidum. (Annunciation Lily). The well known white lily; delightfully fragrant. 20 cents each, $2 per dozen.

Candidum Flore Plena. Same as above, but double. 40 cents each, $4 per dozen.

Harrisii. (Bermuda Easter Lily). Large, pure white; free bloomer. 25 to 50 cents.

Hansonii. Bright yellow, with crimson spots. $1 each.

Longiflorum. Snow white and fragrant. 25 cts. each, $2.50 per dozen.

Thunbergianum. Scarlet. 15 cents each, $1.50 per dozen.

Tigrinum. (Tiger Lily). Salmon, spotted with black; single. 10 cents each, $1 per dozen.

Tigrinum Flore Plena. (Double Tiger Lily). Very hardy; a good bloomer; flowers orange-red, double. 20 cents each, $2 per dozen.

LEONOTIS LEONORUS.

An old plant recently reintroduced, producing long terminal spikes of beautiful orange colored flowers; treated in the same manner as Chrysanthemums, they will flower continually from the early part of September until mid-winter.

MESEMBRYANTHEMUMS.

(Ice Plant.)

Well adapted for baskets, vases, rock work, or for bedding out as borders to flower beds, etc. Its succulent character enables it to stand the hot sun admirably. 10 cents.

MUSA ENSETE.

(Abysinnia Banana.)

The most beautiful of the genus; leaves very large, erect, with a pink mid-rib. A grand plant for a lawn. $1 to $3 each.

MIMULUS MOSCHATUS.

The old well known Musk plant; much admired by some people. 10 cents.

MAURANDYA.

An exquisite slender climbing plant, with graceful foliage and handsome purple trumpet-shaped flowers. 15 cents.

OLEANDER.

Well known; grows and blooms well out of doors in this latitude in summer, but should be taken up in the fall and kept in the house or light cellar. It will survive the winter with but slight protection 200 miles south and east of Fort Worth.

Large fine plants, 75 cents to $1 each; small plants, 25 cents each.

Splendens. Double pink; the best of its color, and very fragrant.
Single White. The hardiest and best bloomer.

MADEIRA VINE.

A very rapid climbing plant, with thick glossy green foliage and fine white fragrant flowers. Winter the tubers in a cellar, same as Dahlias. 10 cents each, 60 cents per dozen.

OTHONNA CRASSIFOLIA.

In habit this somewhat resembles some of our varieties of Sedums. It is excellently adapted for carpeting the ground under shrubs, or as a plant for baskets or vases. It has small, bright yellow, tassel-like flowers, which are borne in great profusion. 10 cents.

PEPINO OR MELON SHRUB.

A shrubby plant of the Solanum family, recently introduced into the United States from Central America; a promising addition to our tropical fruit bearing plants, bearing fruit the size of a goose egg and much the same shape; of a pale lemon color. Interior of fruit a solid pulp, similar to that of a pear and of a taste resembling that of a fine Muskmelon, with a charming acid flavor, allaying thirst in warm weather; delicious and wholesome. The plant is an enormous yielder, commencing to set fruit four to six weeks after being set out, yielding until frost; grows two to three feet high; is hardy in the Southern states with a slight protection of roots with straw or evergreen boughs—said to be quite as hardy as the fig and valuable for the South. We offer small, well-rooted pot plants, ready for immediate planting at 20 cents each, $1.50 per dozen.

PANDANUS.

(Screw Pine.)

Elegant decorative plants of the Palm order, which are extra fine for house use.

Utilis. (Screw Pine.) So called from the arrangement of the leaves on the stem. It is a beautiful plant, excellently adapted for the centers of vases or baskets, or grown as specimens, no plant is better for room culture. 50 cents to $2 each.

Veitchii. This splendid variegated Screw Pine is one of the most attractive plants. The leaves are light green, beautifully marked with broad stripes and bands of white; gracefully curved. $1.

PASSIFLORA CONSTANCE ELLIOTT

This has become a very popular climbing plant. It will live from year to year in the open ground. The flowers are pure white, excepting a very slight coloring at the base of the corolla. To the list of cut flowers and climbing vines it is a decided acquisition. The flowers are of exceedingly attractive form, and produced freely. The plant dies to the ground in the fall, but springs up with renewed vigor in spring, and grows freely. Rarely has as fine a climber as this been offered.

Passiflora Constance Elliott.

PLUMBAGO.

Capensis. A most beautiful plant, producing freely throughout the summer and fall, large trusses of azure-blue flowers, forming a pleasing contrast to the numerous scarlet bedders; of neat and bushy habit, and the plants can be trimmed into very symmetrical shape; does well also in pots as a decorative plant. 15 cents.

Alba. Pure white; flowers in long racemes; very long. 15 cents.

PILEA ARBOREA.

A very neat and pretty plant, resembling some of the ferns in general appearance; its graceful habit makes it desirable for baskets and vases. 10 cents.

PAEONIES, HARDY HERBACEOUS.

A well known genus of plants, noted for their hardiness, ease of culture, vigorous growth in any garden soil, and for the wonderful size and attractiveness of their flowers, which in many sorts are nearly half a foot in diameter, well rounded and perfectly double; different colors. 50 cents each.

Pæonies, Hardy Herbaceous.

PANSIES.

Pansies are of easy culture. They do best in a rich soil and partial shade, though they grow readily in almost any situation, blooming freely all through the spring and early summer. The variety and beauty of their flowers are too well known to need any description. The plants we offer are from the best imported German and English seeds, and cannot fail to give satisfaction. 60 cents per dozen.

PHLOX DRUMMONDII.

Remarkable for the brilliancy and abundance of their large terminal flowers, completely hiding the foliage,; the blooms are of many colors, from pure white to deepest purple, eyed and striped. For masses of separate colors and for cutting for bouquets, they are unsurpassed. White, rose, scarlet, deep blood or mixed colors. Give good, rich ground, and set plants six inches apart each way. 50 cents per dozen.

PHLOXES, PERENNIAL.

Few plants give greater satisfaction to the amateur than the Phlox. The ease with which they are cultivated, their entire hardiness, their extended season of blooming, and the variety and beautiful colors of the flowers, make them very desirable. They succeed well in any good, rich soil, not over-dry. This collection embraces every color from purest white to darkest crimson. 15 cents each, $1.50 per dozen.

Crozy Fils. Rosy salmon.
Esais Tegner. Magnificent rose, purple center.
Foutcheon. Reddish violet, very large.
Gen. Marguerite. Clear lilac, center white, edge of petals white.
J. C. Hanisch.
Lucie Baltet. Rich lilac, edged white, large white center; very fine. 25 cents.
La Reve. Deep rose striped with white.
Lady Musgrove. Pure white, with light purple eye.
Princess Louise. White, with very small pink eye.
White Queen. White.

Perennial Phlox.

PEPEROMIA.

Pretty, dwarf growing foliage plants. They require partial shade and moisture. Good for baskets, etc.

Resedaflora. Leaves small, flowers white, a constant bloomer; very handsome. 25 cents.
Prostata. A very ornamental basket plant, with slender creeping stems, and alternate round leaves. 20cts.

SALVIAS.

Very popular bedding plants, blooming from July until cut down by frost; no other blooming plant affords a more brilliant coloring.

Marmorata Nana. A dwarf grower and very profuse bloomer; flowers beautifully marbled scarlet and white. 10 cents.

Splendens. Covered in autumn with spikes of dazzling scarlet flowers. 10 cents.

Splendens Alba. A pure white flowered variety of the above. 10 cents.

Pitcherii. New; blue, dwarf habit; one of the prettiest and hardiest. 15 cents.

SMILAX.

A popular and well known climber with beautiful foliage of dark glossy green, used largely with cut flowers, etc. 10 cents.

STOCKS, GERMAN TEN WEEKS.

The Stock has for many years been a general favorite. The double flowers are of great fragrance and beauty. Plants of the best double varieties, 10 cts. each, $1 per doz.

SAXIFRAGA SARMENTOSA.

A handsome plant of low habit; leaves nearly round, and striped freely with silvery bands; blooms white, and borne in spikes of nearly twelve inches high; for hanging baskets, vases, etc. This is also known as Strawberry Geranium and Beefsteak Plant. 10 cts.

STEVIA SERRATA.

A very desirable winter blooming variety, never exceeding eighteen inches in height; pure white flowers, fine for cutting. 10 cents each, $1 per dozen.

SOUTHERN WOOD.

The old favorite English sweet-scented plant. 15 cts. each, two for 25 c.

SOLANUM JASMINOIDES.

A beautiful climber for the house in winter, requiring but little care and producing clusters of white flowers. 10 cts.

SANSEVERA ZEYLANICA.

Sword-like leaves; green, marbled with lighter shade. 50 cents.

TUBEROSES.

Everybody in Texas may have Tuberoses for the mere planting of them, if they have large sound bulbs to plant and rich soil in which to grow them.

Dwarf Pearl. It is more double than the common variety, and is of dwarfish habit. Flowers in great profusion and very sweet. 10 cents each, 75 cents per dozen.

Early Single. An early bloomer, and deliciously fragrant. 10 cents.

Variegated. A handsome variegated foliage single variety. Leaves beautifully margined and striped with creamy white and green. 10 cents.

Double Italian. A tall growing, very double and fragrant variety. 10 cts.

Pearl Tuberose.

Verbena.

GRADESCANTIA.

The following varieties of Tradescantia (often called Wandering Jew) have beautifully marked foliage, and are fine for hanging baskets and vases, or for house culture, as they will endure almost any hardship if liberally supplied with water.

Zebrina. Leaves dark green, with a silvery stripe. 10 cents.

Multicolor. Beautifully striped with white, crimson and olive green; sometimes sports. 10 cents.

VERBENAS.

The Verbena is too well known to need any recommendation. It commences to flower and spread from the first day the plants are set, until late in the autumn, every day becoming better and handsomer. If Verbenas are pegged down to the ground as they grow, the plants will extend rapidly, and afford a much greater amount of bloom than if allowed to grow up, when they become "straggling." There are several hundred varieties in cultivation, from which we have selected the following list as being vigorous growers and free bloomers. This list contains the cream of the old varieties, the Mammoth strain (of which the individual flowers are each as large as a twenty-five cent piece), and the latest introductions.

10 cents each, 75 cents per dozen, $5 per 100.

Antler. Ruby.

Beulah. Deep pink, with small white eye.

Bluebird. Brilliant blue, producing large truss. We recommend this variety to any one desiring a blue bed or border.

Beauty of Oxford. Fine large pink.

Brilliant de Vase. Dark crimson, yellow eye.

Champion. Bright crimson, white eye; vigorous and free flowering.

Freddie. Large rosy pink; fine.

Grace Darling. Violet purple; beautiful.

New Mammoth Verbena.

Harlequin. White, flaked with rosy pink.
Jersey Lily. Pure white.
John Brown. Fine dark purple.
Lord Craven. Deep crimson.
Mattie. Shell pink.
Miss Arthur. Dazzling scarlet.
Mayflower. Soft glowing pink; an excellent bedding variety.
Mrs. Woodruff. Bright scarlet.
Mrs. Buchanan. Pink, shaded center.
Maculata. White, striped scarlet.
Negro. Black.
Silver Plume. Fine white bedder.
Snowflake. White.
Vesta. One of the purest whites.
Zenobia. Dark purple, white eye.

NEW VERBENAS. MAMMOTH STRAIN.

10 cents each, $1 per dozen.

Century. Brilliant scarlet, clear white center.
Crystal. Pure white.
Dawson. Rich purple mauve, clear white center.
Edith. Salmon, shaded carmine, center white.
Emily. Royal purple, clear large white center.
Glow-Worm. Brilliant scarlet, perfect form.
Jean. Rosy pink, distinct white center.
Lapiz-Lazuli. Blue, perfect form.
Miss Stout. Carmine scarlet.
Maltese. Lilac, shaded blue.
Mrs. Massey. Salmon pink, large white center.

VIOLETS.

These lovely and sweet-scented flowers are most beaut ful, and the varieties now offered leave little to be desired.

15 cents each.

Marie Louise. Double; deep violet blue; very fragrant and free flowering.
Swanley White. A sport from the popular and well known *Marie Louise;* similar in habit and freedom of flowering, but of a pure white color.
White Czar. Very large, single; white, sweet-scented.

Fruit Department.

PEACHES.

Prices, except where noted, 20 cents each, $2 per dozen, $12 per 100.

Amelia. Very large, conical; white, nearly covered with crimson; juicy and melting. Ripens July 1st to 10th.

Albert Sidney. Medium, oblong; yellowish white, with red cheek; flesh melting and of the highest flavor. July.

Alexander. Above medium, highly colored; flesh greenish white, very juicy, vinous and of good quality. Matures from May 25th to June 10th in Fort Worth. ☞ Many varieties have been offered as being earlier or larger than the *Alexander*, but so far none have proven superior to it.

Amsden. Same as *Alexander*. Tree an upright grower and sure bearer.

Arkansas Traveler. Creamy white, nearly covered with red. Said to be earlier than *Alexander*.

Beatrice. Small to medium; deep red and mottled deeper red; flesh juicy, vinous and of good quality. Ripens after *Alexander* and before *Hale's Early*.

Bonanza. Large to very large; white, with red cheek; fine, very productive. September 1st to 15th.

Columbia. Very large; skin downy, dingy yellow; juicy and rich. Ripe about July 20th, and continues for a month.

Cora. Above medium; white with a pale red cheek; flesh white, juicy and well flavored. Middle to end of September.

Crawford's Early. Large; yellow, with red cheek; flesh yellow, juicy and rich; very productive; a standard market variety. July 1st to 10th.

Crawford's Late. Large; yellow with dark mottled red cheek. One of the very best market peaches. July 20th to 30th.

Croft's Golden. (Cling.) Very large; deep yellow with crimson cheek; flesh yellow, sub-acid. August 1st to 10th.

Chinese Cling. Very large, specimens often measuring fourteen inches in circumference. Usually a shy bearer, but of superior richness and flavor.

Duff's Yellow. (Cling.) Very large; yellow, with red wash; juicy, sub-acid; showy fruit. July 10th.

Early Rivers. Pale straw, with a delicate pink cheek; one of our finest peaches for home use, but too tender to ship long distances; large, delicious and prolific. June 10th to 20th.

Elberta. Large; yellow, with red cheek; juicy and high flavored; flesh yellow. An excellent shipper. Middle of July.

Early York. Large; white, with red cheek, striped with red. Productive; flesh white, melting and juicy. July 10th.

Fleitas or Yellow St. John. Large, roundish; orange yellow, with a deep red cheek; juicy, sweet and highly flavored. June 15th to 25th.

Foster. Large; highly recommended; much like *Early Crawford*, of which variety it is supposed to be a seedling. Ripens ten days before the *Early Crawford*.

Gov. Garland. Fruit medium, handsome; striped or marbled with yellow; flesh thick, white, fine grained, juicy, tender, melting, sweet; very productive. Ripens with *Alexander*.

Great Eastern. Very large, often measuring 14 inches; greenish white with a slight wash of red; flesh juicy and and sweet, sometimes a little coarse; a showy fruit. July 20th. 50 cents each, $5 per dozen.

Honest John. Large; yellow, with red cheek; an excellent peach, of fine flavor; freestone. A good market sort. July 10th to 20th.

Honey. Medium, oblong, with a sharp, recurved point; creamy white, washed and mottled carmine; flesh of a peculiarly fine texture and a honey sweetness; tree very thrifty. A distinct grower and prolific. Ripe about June 25th.

Hale's Early. Of medium size, with red cheek; flesh white, mėlting, vinous, and very good. One of the best bearers, seldom missing a crop. June 5th to 15th.

Henrietta. Large; yellow, crimson cheek; showy; very productive. Clingstone.

Heath Cling. Large; skin creamy white, very seldom any red; flesh pure white to the stone, juicy, sweet, and of good aroma; popular for preserving. Ripens about the last of August.

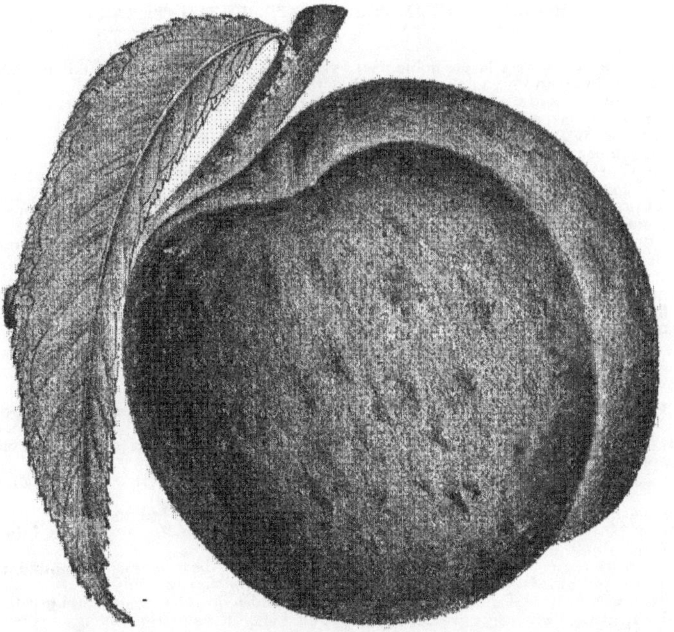

Schumacher Peach.

Indian. Large; dark claret, with deep red veins, downy; flesh deep red, very juicy; vinous and refreshing. Middle of August.

Jacques' Rareripe. Superb yellow; productive, ripening with *Crawford's Late.* Freestone.

Lord Palmerston. Very large; beautiful creamy white, with a blush cheek; of rich flavor and firm, but melting flesh. August 20th.

Leopold. (Cling.) Large, yellow; of fine flavor and very productive. August 1st to 10th. 50 cents each, $5 per dozen.

Mountain Rose. Fruit large; skin white, nearly covered with brilliant crimson; flesh white, melting, sweet, and delicious flavor. Ripe about June 25th. One of the most valuable of early varieties.

Muscogee. Size large; skin dingy yellow, nearly covered with crimson, with a red and dark brown cheek, spotted and somewhat striped like the *Columbia;* flesh white, with some red veins around the stone. Melting, juicy and very good. Beginning of August.

Newington Cling. Large, oblong; white, slightly tinged with red and red cheek; flesh firm, juicy and well flavored. August 10th.

Old Mixon Free. Large, nearly round; creamy white, with red cheek; juicy, sweet and well flavored. Last of July.

Old Mixon Cling. The most profitable clingstone grown at this time of ripening. Fruit, large oblong; a wonderful bearer. July 24th to August 5th.

Osceola. Large; golden yellow, with orange cheek and a few red veins; flesh golden yellow, sweet, buttery and with an apricot flavor. Beginning of September. Freestone.

Pine Apple. Skin golden yellow, tinged with dark red; flesh yellow, slightly red at the stone, juicy, sub acid, excellent. August. 50 cents.

Picquet's Late. Very large; yellow, with a red cheek; flesh yellow, buttery, rich, sweet and of the highest flavor. Maturing from end of August to middle of September. Freestone.

Stonewall Jackson. A seedling of the *Chinese Cling.* Flesh melting, juicy and of high flavor; oblong; creamy white, with crimson wash. Clingstone. July.

Steadley. Large, greenish white; of good quality and productive. August 25th. Freestone.

Susquehanna. A very large, noble fruit; skin deep yellow, with bright red cheek; flesh yellow, juicy and high flavored. End of July.
Schumacher. A very early peach of the *Alexander* type, and closely resembles it. Medium to large; bright yellow, with crimson cheek.
Stump the World. Large; white, with bright red cheek; flesh white, juicy and of good flavor; stands carriage well, and is a fine market variety. July 20th.
Sallie Worrell. Flesh white, juicy and of a delicious flavor; tree a vigorous grower and good bearer. A fine market sort; freestone. September.
Tuskina. Above medium, oblong; skin yellow and deep orange red; flesh sub-acid, vinous and good. Clingstone.
Thurber. A seedling of the *Chinese Cling*, Originated by Mr. P. J. Berckmans of Georgia, who claims for it all the good qualities of the parent with the additional merit of being a freestone of large size. End of July.
Wager. Very large; yellow, more or less colored on the sunny side; juicy and of fine flavor. July.
Waterloo. Medium, red; very sweet. First of June.

APPLES.

Except where noted, 20 cents each, $2 per dozen, $12 per 100.

Early Harvest. Of good size, bright yellow; tender, juicy, well flavored; a fine market apple. Ripens June 10th and lasts two weeks.
Red Astrachan. Large, roundish; nearly covered with deep crimson, having a thick bloom like a plum; juicy, rich, acid. June.
Red June. Deep red, juicy and very productive. Ripe June 15th to July.
Yellow Horse. Large, yellow, acid; fine for cooking and drying. Ripe June and July.
Summer Queen. Large, striped, excellent, productive; bears young and abundantly.
Summer Pearmain. Dull red; delicious, good bearer, hardy July.
Maiden's Blush. Of medium size, clear yellow and red; juicy, tender and good.
Fall Pippin. Very large, greenish yellow, delicious. First of September.
Twenty Ounce Pippin. Very large, yellow, showy. Middle of September.
Jonathan. Medium, red; very beautiful and excellent; productive. October.
Ben Davis. Large, striped, showy; most superb. Flavor moderately good; firm keeper.
Kentucky Red. Medium to large, handsome; good flavor.
Yellow Belleflower. Large, delicious, popular; tree vigorous, spreading, hardy; a fine market fruit.
Smith's Cider. Medium, pale striped; flavor moderate.
Winesap. Medium, red, excellent; hardy, early and a most prolific bearer.
Golden Pippin. Golden yellow; crisp, juicy, tender and of fine flavor. Keeps well.
Gravenstein. Large, excellent, vigorous; productive, though not a sure bearer.
Shockley. Ripens in October, and has been kept until the following August. Although this apple cannot be classed as first quality, yet it is the most popular winter variety in Texas.
Arkansas Black. A new apple from Arkansas. Large; dark red, nearly black; of fine flavor, and a valuable market and keeping variety; not tested. 50 cents each, $5 per dozen.

CRAB APPLES.

50 cents each, $5 per dozen.

Quaker Beauty. Yellow, sub-acid; bears well in Texas.
Whitney's No. 20. A large new early crab, fine for dessert and cooking.
Hyslop. Large; very handsome and popular; fruit deep red.
Transcendant. Large; very beautiful and popular; growth rapid and irregular; a good bearer and valuable. Fruit yellow and red striped, with dark red cheek in the sun.
Siberian Red and Yellow. Well known. Popular in Texas.
Tetofsky. (Russian Crab.) Medium; whitish yellow with crimson stripes; of juicy tender and pleasant flavor; upright, moderate grower with large leaves; extremely hardy and valuable.

PEARS.

Except where noted, 50 cents each, $5 per dozen.

SUMMER PEARS.

Bartlett. By far the most popular pear in America, and deservedly so for its size, beauty and free bearing qualities. Large, buttery, melting, rich. July and August.
Clapp's Favorite. Large, of very good quality; showy and becoming quite popular; ripens before the *Bartlett*.

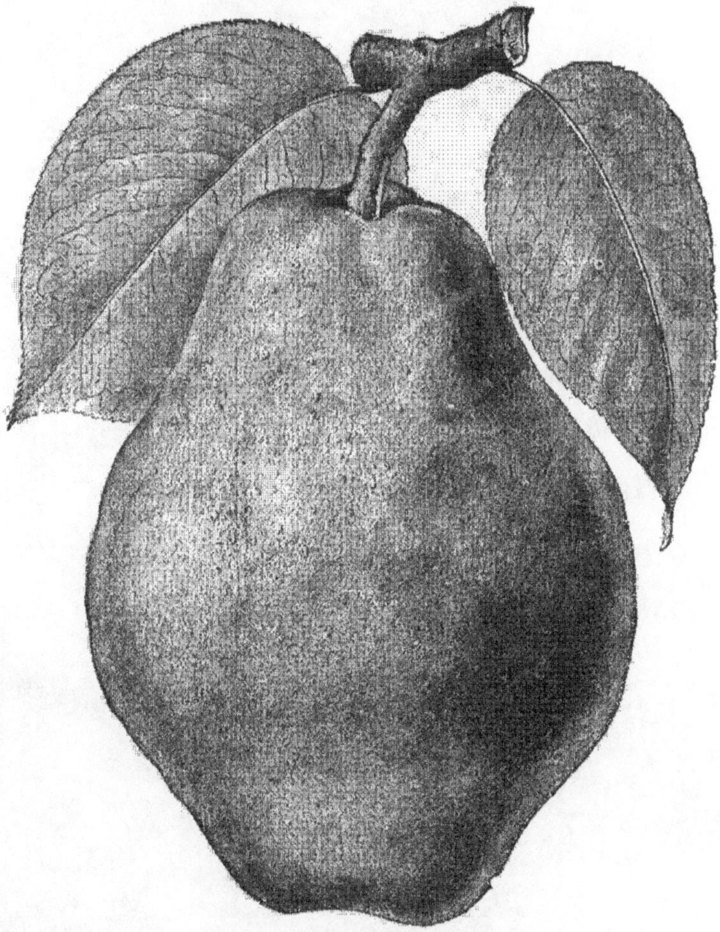

Kieffer Pear.

Doyenne d' Ete. Small, melting, very good; tree a moderate grower. Beginning to middle of June.
Flemish Beauty. Large, melting, sweet, handsome. August.
Howell. Medium; very rich and juicy; a good bearer and fine fruit; tree an open grower. Beginning of August.
Osband's Summer. Small, very good, fine grower.
Seckel. Small but of delicious flavor; tree a stout, slow grower. August.

AUTUMN AND WINTER PEARS.

Kieffer. Origin near Philadelphia, where the original tree, now sixteen years old, has not failed to yield a large crop of fruit for thirteen years past. Tree has large, dark green glossy leaves and is of itself very ornamental; is an early and very prolific bearer. The fruit is of fair quality, wonderfully showy and valuable for canning or market. This and the *Le Conte* will come into bearing two or three years earlier than than the old sorts. Four to five feet, 50 cents each, $5 per dozen; six feet and upwards, 75 cents each, $7.50 per dozen.

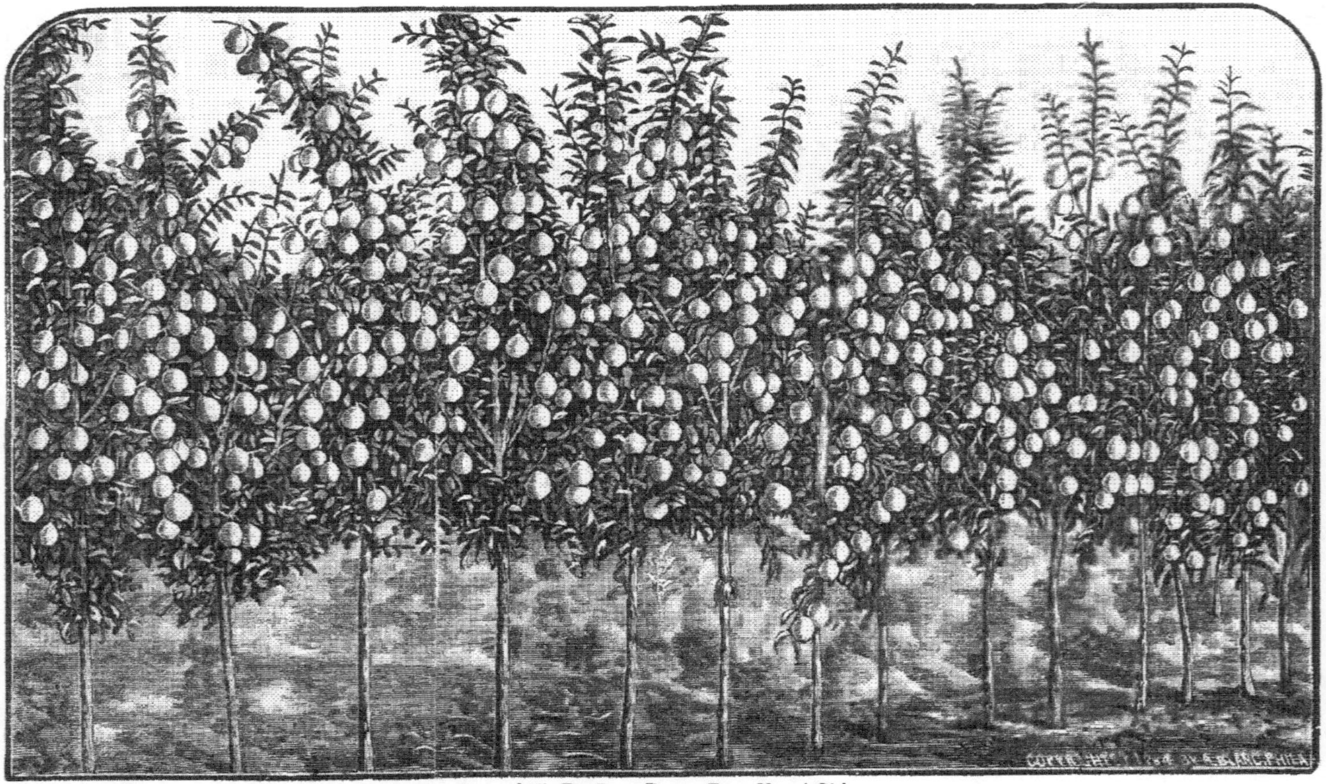

Row of Kieffer Pear Trees, Four Years Old.

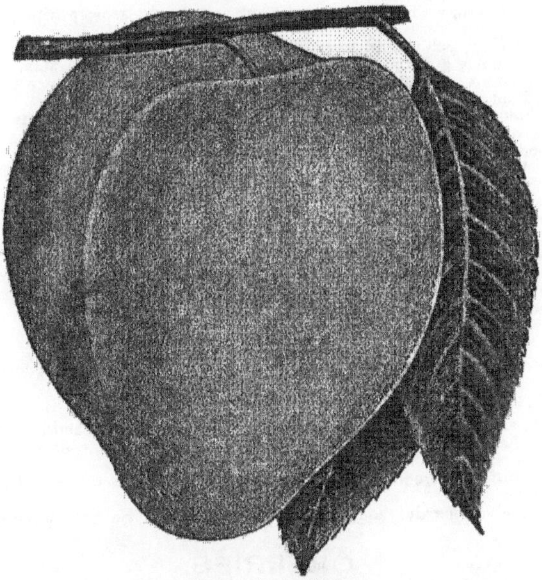

Kelsey's Japan Plum.

Beurre d' Anjou. Large; juicy, melting, sometimes a little astringent; fine tree and regular bearer. August to September.

Duchesse d' Angouleme. Very large, melting, jucy and well flavored; sometimes weighs over a pound. Greenish yellow, with some russet; does best on quince. Middle of September.

Le Conte. Fruit large, pyriform; skin smooth, pale yellow; quality very variable—usually of second quality, but if allowed to mature slowly in a cool, dark room or in drawers, its quality improves remarkably. Maturity from first to end of August. Trees begin to bear when four years old, and should be planted twenty feet apart each way. Four to five feet, 50 cents each, $5 per dozen; six feet and upwards, branched, 75 cents each.

PLUMS.

Except where noted, 25 cents each, $2.50 per dozen.

Caddo Chief. Earliest; fruit round, red; tree prolific, holds its foliage late, and therefore should be transplanted in winter or spring. From Louisiana.

Clinton. Similar to Weaver. August.

De Soto. Large, round, dark purple; one of the best; prolific.

De Caradene. Red, round, acid. June.

Blackman. Tree resembles the peach; a fine grower, but a shy bearer.

Golden Beauty. Medium, yellow, round. From Western Texas. August.

KELSEY'S JAPAN. This plum is now conceded by all who have tried it to be the best late plum in America. Our trees fruited last season, and we can freely confirm all that is said in its praise. It has been in bearing in the United States since 1876 and the trees have never failed to produce all the fruit they could carry. The following points are claimed for it: "Its wonderful productiveness is not surpassed by any other plum, native or foreign. It comes into bearing at the age of three years. The fruit is a very large size, being from seven to nine inches in circumference, with a small pit; specimens sometimes weigh six and a-half ounces each. Color rich yellow, nearly overspread with bright red, with a lovely bloom. It is of excellent quality, melting, rich and juicy. Its large size renders the parting of the fruit as practicable as the peach, and it excels all other plums for canning. As a dried fruit this is destined to take the lead, being equal to, if not surpassing the best dried prunes. In texture it is firm and meaty, possessing superior shipping qualities. It ripens from the first to last of September." 50 cents each, $5 per dozen.

Green Gage. Small, well liked where known; tree a good grower with us. Middle of July.

Lone Star. Large; yellow and purple; rich, melting and sweet; the best early plum; of dwarf, weeping growth; should be transplanted late. Last of May.

Marianna. Tree a vigorous handsome grower; fruit, medium, red, sweet.

Miner. Medium, dark red; a good grower.

Newman. Medium, oblong; skin a beautiful glossy red color, with delicate purple bloom.

Prunus Simonii. Comes highly recommended, though it has not yet fruited with us. Fruit good size, brick red, with a fine apricot flavor. 50 cents.

Robinson. We have tested this sufficiently to warrant us in saying it is equal to *Wild Goose* in every respect, with the difference that it ripens two weeks later, thus making it a most valuable market variety. With a good orchard of *Caddo Chief*, *Wild Goose*, *Robinson* and *Kelsey*, you can supply the market with fine plums nearly all summer. 50 cents each, $5 per dozen.

Weaver. Large, red; tree a good grower, but we fear a shy bearer.

Wild Goose. The "old reliable." Too well known to need description. It stands among plums where the *Bartlett* does among pears and the *Concord* among grapes. No collection is complete without it.

APRICOTS.

25 cents each, $2.50 per dozen.

Early Golden. Medium, pale orange; flesh yellow, juicy, sweet and good.

Moorpark. Large size; orange, brownish red in sun; flesh quite firm; juicy, with rich luscious flavor.

California Black. More like a plum, and is probably a cross between the Apricot and Plum. June.

Russian. Tree very hardy, though dwarfish in its habits. Not yet fruited with us.

CHERRIES.

Black Tartarian. Very large; productive where it does well. May 15th. 50 cents.

Coe's Transparent. Medium sized, mottled. May 20th. 50 cents.

Olivet. A new French variety. Fruit large and globular, with a very shining deep red color; flesh red, tender and vinous. 75 cents.

Governor Wood. Large, yellow and red; excellent. June 1st. 50 cents.

Early Richmond. Medium, red, early; hardy and productive; one of the best cherries for the south. 50 cents.

May Duke. Large, dark red; tree hardy, bears well, and is quite distinct. 50 cents.

Utah Hybrid. Fruit medium to large, rich reddish purple, and of uniform shape and size. As a table fruit, only fair in quality. Tree dwarfish; a sure and prolific bearer. A good market variety. 50 cents.

QUINCES.

Orange or Apple. Large, orange shaped and of excellent flavor. 50 cents.

Champion. The tree is a strong, free grower, more like the apple than the quince, and usually comes into bearing the second or third year. Very productive and of largest size. $1 each, $10 per dozen.

NECTARINES.

50 cents each.

NUT-BEARING TREES.

Almonds, Soft-Shell—Princess and Sultana. Both are prolific, soft-shelled and good. These are the varieties mostly cultivated in Europe, and produce the bulk of the almonds of commerce. The trees are rapid, upright growers, and highly ornamental. 50 cents each, $5 per dozen.

Almond, Hard-Shell. Fruit large; more hardy than the soft-shell. Not so showy, but still quite ornamental. 50 cents each.

FIGS.

50 cents each, $5 per dozen.

Will easily produce a full crop in Texas, and are worthy attention from our ruit growers.

Japan Persimmon.

JAPAN PERSIMMONS.

Too thoroughly tested to be regarded as a novelty now. It has proved a success wherever cotton will grow, and no collection of trees, whether fruit or ornamental, should be without it. Home grown trees, $1 each.

GRAPES.

Brighton. Vine hardy, a rapid and vigorous grower; leaves large, thick, dark; quality best as compared with the finest table grapes: the fruit ripens early, along with the *Hartford Prolific* and before *Delaware;* is good as soon as fairly colored red, becoming very rich, but less vinous if allowed to hang long. 50 cents each, $5 per dozen.

Agawam. Bunches medium to large, often shouldered; berries very large, pulp soft, sweet, sprightly, of peculiar aromatic flavor; productive and of great vigor of growth; prefers long pruning. Ripens soon after the *Concord.* 35 cents each, $3.50 per dozen.

Concord. Bunches large; berries very large, blue black, with bloom; skin thin, pulp dissolving, juicy; a beautiful market variety; rampant grower and good bearer. Ripe middle of July. 15 cents each, $1.50 per dozen.

Delaware. Bunches medium, berries medium to small, with red or pink skin, very thin; juicy, vinous and most delicate table grape; a very prolific bearer. 35 cents each, $3.50 per dozen.

Catawba. The best late grape for this section. Bunches and berries of good size; fine coppery or purplish red, and when well ripened of delicious flavor; vigorous, early and a grand bearer.' One to two weeks later than *Concord.* 25 cents each, $2.50 per dozen.

Champion. Said to be the earliest black grape; bunches and berries large, black; quality good; vine hardy, vigorous, productive. Valuable for market. 25 cents each, $2.50 per dozen.

Black Spanish. (*Lenoir.*) Of the Herbemont class. Bunches medium to large, shouldered under favorable circumstances; on badly pruned or overloaded vines the bunches are loose and not shouldered; berries small, round; vine a fine grower and abundant bearer. 25 cents each, $2.50 per dozen.

El Paso or **Mission.** Bunches very large, often weighing one and a-half pounds; berries medium to small, of dark amber color; flavor the best; commands fancy prices on any market. Not quite hardy here; needs slight protection in winter. If well cared for, it will repay any one for the trouble. We consider it the best grape for Western, South-Western and Southern Texas. 50 cents each, $5 per dozen.

Hartford Prolific. Bunches large; berries large, blue; flesh pulpy, musky, sweet. July 1st. 25 cents each, $2 50 per dozen.

Rulander. Small, compact; purple, sweet, fine quality; an excellent wine grape. 25 cents each, $2.50 per dozen.

Scuppernong. As this is such a general favorite of all who are familar with it, we grow and sell a few. It has never proved a success from Central Texas, north and west, though succeeding well in the southeastern portion of the State. 50 cents each.

Triumph. Bunches and berries very large, white; vinous, excellent; a most showy variety; productive and perfectly free from rot. Becoming very popular. 75 cts. each.

Herbement. Bunch large, long, shouldered and compact; berries small, black, sweet, juicy, highly flavored; a fine wine grape. 35 cents each, $3.50 per dozen.

Goethe. (*Rogers' No. 1.*) Bunches and berries large, greenish yellow, turning to pink at full maturity; very sweet, and of well defined aroma. August. 25 cents each, $2.50 per dozen.

Ives' Seedling. Bunches large; berries large, blue, skin thick; flesh pulpy, musky, sweet; vigorous grower and prolific bearer. Ripens early; an excellent wine grape. 15 cents each, $1.50 per dozen.

Lindley. (*Rogers' No. 9.*) Bunches and berries large, red, and fine flavored. July to August. 35 cents each, $3.50 per dozen.

Niagara. The famous new white grape. A cross between the *Concord* and *Cassady* Bunches large; berries larger than the *Concord*, mostly round, light greenish white, slightly ambered in the sun; skin thick but tough, and does not crack; has a flavor and aroma peculiarly its own, much liked by most people; a fine market variety Like all new fruits, the *Niagara* has been critically watched by all who are in any way interested in grapes, and no one has watched it more closely than we—if it proved to be as represented by its introducers, "the grape for the million," we wanted to know it; and if a failure, we did not care to push the sale of it. Our neighbor, Mr. W. F. Blandon, has about 2,600 vines of the *Niagara*, which he bought from the Niagara White Grape Company. The vines were loaded. At the time the *Niagara* came on the market, the price of grapes here had run down to five cents per pound, and slow also at that. The *Niagara* readily brought 12½ cents and the whole crop was closed out at that price, and was retailed at 15 and 20 cents a pound, where *Concord* and other varieties went begging at 6 to 8 cents. We would advise all who want a good white grape that will be sure to bear a heavy crop of fruit to try a few *Niagaras*. Good for trellises or arbor. July and August. 75 cents each, $8 per dozen.

☞ *We are growing other varieties of grapes and will quote in our catalogue from time to time, such as prove worthy.*

BLACKBERRIES.

Dallas. This is described by Mr. Howell, of Dallas, as follows: "A native, discovered in Dallas several years ago. By culture it has been greatly improved in size, flavor and productiveness, and to-day it stands at the head of the list for earliness, productiveness and absolute freedom from rust. The Central Texas Horticultural Society, at its meeting in Dallas on the 6th day of August, 1884, adopted a resolution offered by J. M. Howell to name this berry the Dallas Blackberry." 15 cents each, $1.50 per dozen, $4 per 100.

Early Harvest. A new variety from Southern Illinois. The bush is an upright grower; canes strong, perfectly hardy, very productive; berries small and of good quality; very early, ripening several days before the *Wilson's Early*. 10 cents each, $1 per dozen, $4 per 100.

Kittatiny. Very large, sweet and productive, where not affected by rust. This and the *Dallas* give a succession of fruit for six weeks or two months.

STRAWBERRIES.

25 cents per dozen, 75 cents per 100.

Crescent Seedling, Wilson's Albany, and **Chas. Downing.**

RASPBERRIES.

Raspberries succeed well in most part of Northern Texas. The Black-Caps stand the hot, dry summer better than the Red-Caps. They should be planted in rich soil and well worked. We can furnish the *Gregg*, *Mammoth Cluster* and *Doolittle*, black; and *Turner* and *Philadelphia*, red. $1.50 per dozen.

GOOSEBERRIES.

This fruit prefers a deep, rich, rather cool soil. The plants should be annually pruned, and if properly treated will produce abundantly.

15 cents each, $1.50 per dozen.

Houghton's Seedling. An American seedling from Massachusetts. Vigorous, very productive, and free from mildew; pale red when ripe; of medium size and fine flavor.

Downing. Of large size and fine quality; oval, greenish white; plant very productive.

CURRANTS.

Except where noted, 15 cents each, $1.50 per dozen.

Cherry. Very large, deep red, acid; growth stout and vigorous.
Red Dutch. Medium size, deep red; of rich acid flavor; very productive and reliable.
Fay's Prolific. New; very large, deep red. Said to be the best variety grown; untried with us. 50 cents each.

ASPARAGUS.

Conover's Colossal. A mammoth variety of vigorous growth, which we think the best for this climate. 50 cents per dozen, $1.50 per 100.

RHUBARB OR PIE PLANT.

This deserves to be ranked among the best early fruits in the garden. It affords the earliest material for pies and tarts, continues long in use, and is valuable for canning. Make the border very rich and deep.

Linnæus. Large, early, tender and fine; very best of all. Good roots, 25 cents each, $3 per dozen.

☞ *We will be glad to give the benefit of our experience in making selections of fruit trees for any purpose, and will be pleased to furnish lists suitable for any location or purpose.*

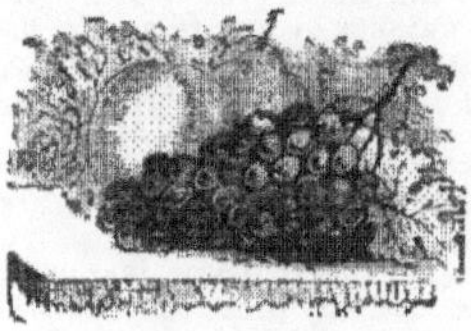

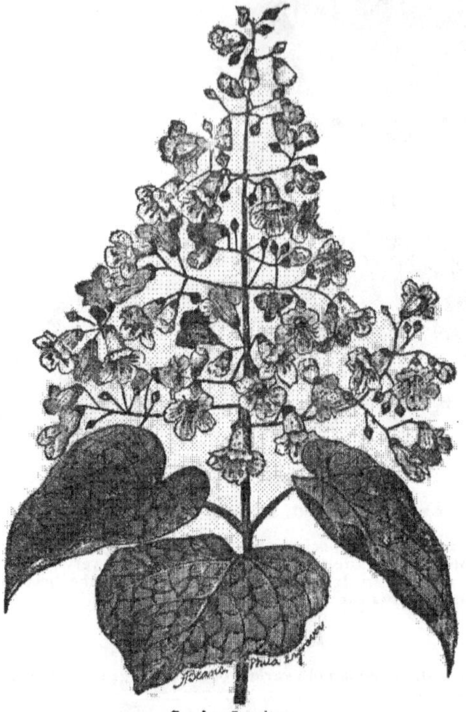

Catalpa Speciosa.

Ornamental Department.

SHADE AND FOREST TREES.

ASH, Native. A well known tree. 50 cents each.

BOX ELDER. (*Ash-leaved Maple.*) Native; a good grower, and makes excellent shade. 50 cents each.

COTTONWOOD. For rapid, handsome growth and hardiness this has few equals. It thrives in any locality—in marshes or on dry sandy hills. Quite desirable for wind-breaks or sidewalks. Trees 8 feet, 25 cents each, $2.50 per dozen; 8 to 10 feet, well branched, 35 cents each, $3.50 per dozen; heavy trees, delivered in Fort Worth only, 50 cents each.

CATALPA SPECIOSA. As we consider the Catalpas among the very hardiest trees for this country, we grow them extensively. Our trees are admired by hundreds of visitors for their uniform beauty, healthfulness and rapid, upright growth. The tree attains sixty to ninety feet in height, and in some instances three to four feet in diameter. The timber is more durable than any other native tree. Several instances are recorded of Catalpa posts having been in use sixty to ninety years and still remaining sound and good. J. C. Teas, of Missouri, says: "Of all the trees that have been suggested as adapted to the formation of timber plantations, the *Catalpa Speciosa* stands pre-eminent. Its exceedingly rapid growth, its adaptation to almost all soils and situations, its wide range of latitude, extending from Canada to the Gulf of Mexico, its almost extraordinary success on the western and northwestern prairies, the ease and certainty with which it is transplanted, its strong vitality and freedom from insects, the incomparable value of its timber for the most important as well as minor uses for which timber is needed, the almost imperishable nature of the wood when used for posts, railroad cross-ties and in other exposed situations, to say nothing of the handsome and stately appearance of the tree, and the unrivaled beauty of its flowers, all point to *Catalpa Speciosa* as the tree to plant."

	Each	Dozen.
Price, transplanted trees, 10 to 12 feet	$0 75	$8 50
" " " 8 feet	50	5 00
" " " 6 "	35	3 50
" " " 4 "	25	2 50
" Seedlings, 10 to 15 inches . . . $1.50 per 100, $10 per 1,000.		

CATALPA Kæmpferi. Very hardy; has fragrant flowers. Transplanted trees same price as *C. Speciosa*; seedlings, 10 to 15 inches inches, $2 per 100, $12 per 1,000.

CATALPA, Tea's Japanese Hybrid. A cross between *Speciosa* and *Kæmpferi*. It partakes of the characteristics of both its parents, having the fragrant, free-flowering habit of *Kæmpferi*, and the rapid upright growth of *Speciosa*. We have the above three varieties growing side by side, and it is difficult to tell which one is the best—they are all wonderful growers and very ornamental. Prices for transplanted trees same as *Speciosa*; seedlings, 12 to 18 inches, $2 per 100, $12 per 1,000. For large quantities, write for special prices.

ELM, White. The Elm is indigenous to every part of Texas, and will do well almost anywhere. Our trees are transplanted and the most of them have nice tops formed. For a handsome long-lived tree, this can be depended upon. 6 to 8 feet, 50 cents each, $5 per dozen; 8 to 10 feet, 75 cents each, $7.50 per dozen.

LOCUST, Black. Quite hardy anywhere in the State. 25 to 50 cents each.

MAPLE, Soft or Silver-Leaved. The very rapid growth, spreading branches and silvery foliage of this well known tree have induced its extensive planting for ornament and shade. It thrives in nearly all soils, and is one of the handsomest and most desirable of shade trees for this climate. 6 to 8 feet, 50 cents each, $5 per dozen; 8 to 10 feet, well branched, 75 cents each, $7.50 per dozen.

MULBERRY. Nearly all mulberries are hardy in Texas. They are rapid growers, long lived and handsome. The following we have found desirable for this locality.

Russian. Dense, dark green foliage; fine for hedges, wind-breaks, forest or shade; the hardiest of all mulberries, and a prolific bearer of fair fruit; is a fine tree to plant in quantity, either for fruit or defense. 5 to 7 feet, 50 cents each; 12 to 18 inches, $2.50 per 100, $15 per 1,000.

Hick's Everbearing. We find this the hardiest everbearing mulberry that we have tried. It is a rapid, upright grower, and a continuous bearer. 25 cents each; 6 to 8 feet, 50 cents each.

Multicaulis. Non-bearing; a beautiful shade tree, and largely grown for silk culture. 50 cents each.

PAULOWNIA Imperialis. Has immense leaves, and bears fragrant purple flowers very early; of rapid growth. $1 to $3 each.

POPLAR. All poplars are hardy, and rapid growers.

Balsam or **Balm of Gilead.** 50 cents each.

Carolina. (See Cottonwood.)

Lombardy. Pyramidal in shape, and a fast grower; the most graceful and stately avenue tree. They are not easily broken by the wind, as many suppose. 6 feet, branched, 25 cents each, $2.50 per dozen; 8 feet, branched, 35 cents each, $3.50 per dozen; 12 to 14 feet, extra, 50 cents each, $5 per dozen.

Silver. Beautiful foliage. 50 cents each.

Bolleana. A Russian silver-leaved Lombardy Poplar, lately introduced. It possesses the beautiful silvery foliage of the Silver Poplar, and the regular upright growth of the Lombardy. The leaves are almost black-green on the upper surface, and white underneath; distinct from all others. $1 each.

SYCAMORE. Universally admired for its rapid, upright growth; fast becoming the popular shade tree of our cities. 4 to 6 feet, 50 cents each; 6 to 8 feet, 75 cents each; 8 to 10 feet, $1 each.

UMBRELLA CHINA. A native of Texas, and peculiarly adapted to our climate. Will make dense serviceable shade quicker than any tree we have. The most symmetrical tree in existence, every head being as perfect as an umbrella. Two year old trees that we set out sixteen months ago have formed heads eight to ten feet in diameter and so dense that the sun's rays never penetrate them. In spring, they are covered with a profusion of flowers as pretty and fragrant as a lilac. Our trees have straight, thick bodies and good roots. 8 to 10 feet, heads formed, $1 each, $10 per dozen; 7 to 8 feet, 75 cents each, $7.50 per dozen; 6 feet, 50 cents each, $5 per dozen; 4 feet, 25 cents each, $2.50 per dozen. Price by the 100 or 1,000 of the different sizes given on application.

MISCELLANEOUS ORNAMENTAL AND FLOWERING TREES.

We have tested many of the ornamental trees of the north, but most of them are unsatisfactory, such as Cut-leaved Mountain Ash, Weeping Ash, Kilmarnock Weeping Willow, Camperdown Elm, Weeping Linden, etc. The trees named in the list below prove to be quite hardy. We hope to be able to add to this list every year.

WILLOW, Babylonian. The well known weeping willow; of rapid growth, hardy and graceful. 50 cents each.
WILLOW, Wisconsin Weeping. Makes a large, beautiful tree; hardy. 50 cts. each.
WILLOW, Annularis. Of rapid growth; erect, and leaves singularly curled like a ring. 75 cents each.
PEACH, Poplar or **Pyramidal.** Grow to a height of twenty feet, in a compact form like the Lombardy Poplar. 50 cents each.
PEACH, Dwarf. Of low, bushy growth; rich, dark green, glossy foliage; fruit large and golden. 50 cents each.
PEACH, Blood-Leaved. Leaves blood red in the spring, changing to purple and finally to brown; fruit white. 50 cents each.
PEACH, Double Flowering White. Flowers as double as a Camellia. The tree is a magnificent sight in spring. 50 cents each.
PEACH, Double Red. A bright red variety of the above. 50 cents each.
PRUNUS Triloba. (Double Flowering Plum.) Of vigorous growth; flowers semi-double, of a delicate pink, upwards of an inch in diameter, thickly set on the slender branches. A choice and very attractive spring blooming plant. 50 cents each.
PRUNUS Pissardii. The most valuable of all purple-leaved trees. It retains its deep color throughout our warmest weather, and its leaves until mid winter. 75 cents each.
POPLARS. See SHADE TREES.
MAGNOLIAS. See EVERGREENS.

EVERGREENS, BROAD-LEAVED AND CONIFEROUS.

A few years ago it was thought that evergreens in Texas, and particularly in West Texas, were a failure, but time has proved quite differently. It is true that such trees as the Pine, Spruce, Larch, Yew, some of the Cypresses and Junipers, do not always give satisfaction to the planter, yet every garden can have the beautiful Arborvitæ, Retinospora, Wild Peach, Red Cedar, Box, Euonymus, and that grandest of all evergreens, the stately Magnolia; while in Southern and Eastern Texas, the Cape Jasmine, Oleander, Pittosporum, etc., grow to perfection.

While evergreens are often transplanted in the fall, yet we find that spring planting is much more successful, and we would advise our friends, particularly those in North Texas, to plant in the spring, say from February 1st to March 15th.

Never expose the roots to the air or sun; plant the dirt *firmly* around the roots, water and mulch, and they will be as sure to grow as peach trees.

ARBORVITÆ, Golden. Beautiful compact tree of a golden hue. The most desirable; will never out-grow its compact form. 12 to 15 inches, 25 cents each; 15 to 20 inches, 50 cents each; larger plants, $1 to $2 each.
ARBORVITÆ, Hybrid Golden. Beautiful; nearly as compact as the Golden, and makes a more rapid growth. 25 to 75 cents each.
ARBORVITÆ, Chinese. Faster grower than the above, but not so compact. Bears pruning well, and can be made a beautiful tree. 25 to 50 cents each.
BOX, Tree and Dwarf. Beautiful for hedges or single specimens. Should be planted in clay or black soil; do not thrive in sand. 25 to 75 cents, according to size.
CUPRESSUS Pyramidalis. Dark green dense foliage; very compact and pyramidal, like the Lombardy Poplar. 50 cents to $1 each.
CUPRESSUS Horizontalis. Rapid growing; branches spreading. 75 cts. to $1 each.
CAPE JASMINE. Universally admired for its beautiful glossy foliage and fragrant white flowers. Perfectly hardy south of here, but needs protection in winter in this latitude. Nice plants, 50 cents each; large, $1 each.
CEDAR, Native Red. Makes a beautiful tree when pruned so as to make a compact growth 50 cents to $1 each.
CEDRUS Libani. (Cedar of Lebanon.) Vigorous, wide spreading horizontal branches, foliage dark green; massive and very picturesque. $1 to $2 each.
CEDRUS Deodara. The great cedar of the Himalayan mountains. Of pyramidal form; foliage light silvery green; a magnificent tree. $1.50 to $3 each, according to size.

LIBOCEDRUS Decurrens. A magnificent rapid growing tree; foliage in form of fans. Attains the height of 50 to 80 feet. $1 to $3 each.

MAGNOLIA Grandiflora. The handsomest of all southern broad-leaved evergreens, and well known. They are indigenous to Southern Texas and do well all over the state. There are a few handsome specimens in Fort Worth that have bloomed two or three seasons. 1 foot, 75 cents each; 2 feet, $1.50; 4 feet, $2.50 each.

MAGNOLIA Fuscata. (Banana Shrub.) A dwarf growing variety; in April and May it is covered with a profusion of small flowers, exhaling a most exquisite fragrance, similar to a ripe banana; a favorite. 8 to 10 inches, bushy, $1; 12 to 15 inches, $1.50.

MAHONIA Aquifolium. A low growing shrub, with purplish green, shining, prickly leaves, and bearing showy, bright yellow flowers in spring. 50 cents each.

OLEANDER. Well known; grows and blooms well out of doors in this latitude in summer, but should be taken up in the fall and kept in the house or a light cellar. It will survive the winter with but slight protection 200 miles south and east of Fort Worth. Large fine plants, 75 cents to $1 each; small plants in spring, 25 cents each.

Splendens. Double pink; the best of its color, and very fragrant.

Single White. The hardiest and best bloomer.

PITTOSPORUM Tobira. Glossy dark green leaves and fragrant cream colored flowers; about as hardy as the Cape Jasmine. 50 cents each.

PITTOSPORUM, Variegated. Leaves green, margined white. 50 cents each.

RETINOSPORA. (Japan Cypress.) A pretty class of small evergreens. Bear transplanting well and are quite hardy. 50 cents to $1.50 each.

Plumosa. A hardy, graceful, rapid grower, with delicate glaucous foliage.

Squarrosa. Another rapid grower; has a round bushy head, covered with numerous small leaves of a whitish green tint; densely branched.

Aurea. Gold tinted; very pretty.

WILD PEACH. (More correctly Evergreen Cherry.) One of the prettiest evergreen trees we have. They have all been transplanted, giving them more roots, so that they will bear transplanting again with little loss. 8 to 10 inches, bushy, 25 cents each.

PRIVET. The Privets are almost evergreen here, particularly the Japan and California, which frequently hold their leaves until February. See SHRUBS.

HEDGE PLANTS.

PRIVET, California. Nearly evergreen; a strong growing pyramidal shrub, with bright green leaves and white flowers. This is the most desirable and beautiful hedge plant, grows rapidly and can be trimmed in any shape. $10 per 100.

PRIVET, Common European. Dark green and smaller leaves than the California.

ARBORVITÆ, Chinese. Evergreen; fast growing variety. Makes a beautiful ornamental hedge when neatly pruned. 18 to 24 inches, $20 per 100.

OSAGE ORANGE. Rapid in growth and very thorny. $7.50 per 100.

PYRACANTHA or Evergreen Thorn. $12.50 per 1,000.

☞ We do not grow the last two named, but can supply them on short notice. Special prices on large quantities.

BOX, Tree and Dwarf. Makes a beautiful evergreen hedge. Will not succeed in sand, but flourishes in clay or black land. 25 cents each, $20 per 100.

ORNAMENTAL GRASSES.

ARUNDO Donax Variegata. (Ribbon Grass.) This scarce and beautiful variety is one of the most stately of silvery variegated reed-like plants, and one that can be used either as a single specimen or in groups, its graceful foliage, creamy white and green striped, contrasting well with other foliage plants. 50 cents each.

PAMPAS GRASS. A most ornamental plant, with silvery plume-like spikes of flowers; very hardy, and thrives in almost any ordinary rich soil. 50 cents each.

ERIANTHUS Ravennæ. Attains the height of ten or twelve feet, throwing up numerous flower spikes of a grayish white color; blooms profusely and remains in bloom a long time; needs space to show its merits. The plumes, like those of the Pampas Grass, make elegant winter decorations. 50 cents each.

EULALIA Japonica. A hardy perennial from Japan, with long narrow leaves, striped with green and white; it throws up stalks four to six feet high, terminating with a cluster of flower spikes, on which the individual flowers are arranged; the flowers are surrounded with long silky threads, which when fully ripe or when placed in a warm room, expand, giving the whole head a most graceful and beautiful appearance, not unlike that of an ostrich feather curled. 50 cents each.

EULALIA Japonica Zebrina. Striking and distinct. Unlike most plants with variegated foliage, the striping or marking is across the leaves instead of longitudinal, the leaves being striped every two or three inches by a band of yellow one-half inch wide. In the fall it is covered with flower spikes similar to that of *E. Japonica.* 50 cents each.

Hydrangea Paniculata Grandiflora.

HARDY FLOWERING SHRUBS.

Except where noted, 25 cents each; large plants, 50 cents each,

ALMOND, Double Flowering White. One of the most beautiful of our early flowering shrubs. 50 cents each.
ALMOND, Double Red. A variety with rose-colored flowers.
ALTHEA. (Rose of Sharon.) We are cultivating several varieties of this beautiful shrub, differing in color and shape of flower. They are hardy. of easy cultivation, and desirable on account of their blooming during the autumn months, when there are but few other flowers.
Double Red.
Double White.
Double Purple.
Elegantissima. Semi-double pink; large flower and very free bloomer.
Variegated. Leaves conspicuously margined creamy white. Holds its color well all through the summer; flowers purple, small. 50 cents each.
BUDLEYEA Lindleyana. Pale blue flowers, borne in long pendant racemes. Good grower and constant bloomer.
CRAPE MYRTLE. Truly a southern shrub, and the most showy summer bloomer we have. So well known as to need no description.
Pink. 35 cents each.
Purple. 35 cents each.
New Crimson. 50 cents each.
White. 75 cents each.
CALYCANTHUS Floridus. (Sweet-scented Shrub.) Leaves soft, downy beneath; flowers fragrant like strawberries, double and of chocolate color. 50 cents each.
CURRANT, Yellow Flowering. Blooms early in spring.

CYDONIA JAPONICA. (Pyrus Japonica or Japan Quince.) Scarlet; an old favorite, having a profusion of bright scarlet flowers in early spring. One of the best early shrubs we have; no flower garden should be without one or more.

DEUTZIA. A class of very useful and popular flowering shrubs, of compact and bushy growth.

Crenata. Double flowering; a handsome variety. Flowers white, tinged rose.

Crenata fl. pl. Alba. New; flowers pure white.

Gracilis. Flowers pure white, bell shaped; dwarf and graceful; is often forced by florists for its pretty flowers, which a,.pear with very little heat in the winter.

Pride of Rochester. Double pink; fine.

HONEYSUCKLE, Tartarian Upright or Bush. A vigorous shrub of upright habit, suited to almost any soil and exposure; leaves good size and rich green; flowers pinkish, coming early, before the leaves, and very fragrant. 35 cents each.

HYDRANGEA. Vigorous, spreading shrubs with large showy leaves and great panicled flowers; well known and valuable. Thrive best if planted in partial shade and somewhat moist ground.

Otaksa. Flowers pink, occasionally blue, in immense trusses. 50 cents each.

Paniculata Grandiflora. Of spreading form, bearing immense pyramidal panicles of white flowers, more than a foot long, which change to pink, and finally to purple. Blooms from June to frost; one of the best ornamental shrubs, and popular everywhere. 50 cents each.

Thomas Hogg. Plant not so strong as above. Abundant white flowers in immense heads, often ten to fifteen inches in diameter; the best of its class, and very fine for pot culture. 50 cents each.

JASMINE, Catalonian. Foliage beautiful; fragrant white flowers. Hardy here.

LILAC. Well known, hardy and fragrant. Grow well in almost any soil and are very beautiful.

Purple. The well-known variety; good bloomer.

White. Rich glossy foliage; flowers pure white. 75 cents each.

Persian Purple. Has a heavy crop of blooms in the spring, and then blooms occasionally through summer and fall. 50 cents each.

PHILADELPHUS. (Syringa or Mock Orange.) Strong growers; beautiful and showy.

Grandiflora. Flowers very large, slightly fragrant. 35 cents each.

Coronarius. White; fragrant. 35 cents each.

Coronarius fl. pl. Semi-double flowers. 35 cents each.

SPIREA. A very extensive class of flowering shrubs, embracing many forms and colors of blooms; popular and good.

Billardii. Large spikes of deep pink flowers; profuse and perpetual bloomer; fragrant.

Callosa. Flowers pink: everblooming.

Callosa Alba. Flowers white; everblooming, of dwarf growth; very neat and des rable. 35 cents each.

Callosa Superba. Flowers pale flesh.

Prunifolia. Flowers small, pure white, very double, produced in great profusion, upon long slender branches; blooms early in spring. 35 cents each.

Reevesii. (Bridal Wreath.) Single white.

Reevesii fl. pl. Large round clusters of double white flowers, covering the whole plant. One of the best; blooms in spring.

Thunbergii. Low growing, with beautiful foliage; abundant white flowers. One of the most charming of all low growing shrubs. 50 cents each.

POMEGRANATE. Flowering and fruiting. Perfectly hardy south of this latitude. Here they are killed to the ground during the coldest winters, but grow out again in the spring.

Double White.

Double Red.

SNOWBALL. A favorite old shrub, producing large globular clusters of white flowers. 50 cents each.

SNOWBERRY. Pink flowers in spring, and quantities of large white waxy berries in autumn.

WIEGELIA. Very useful shrubs, bearing pretty flowers at a time when bloom is scarce.

Amabilis. Of robust growth; pink, very profuse bloomer. Spring and fall.

Amabilis Alba. Flowers white, turning pink soon after opening.

Nivea. Rather dwarfish habit; flowers pure white, in spikes.

Rosea. Fine rose-colored flowers in spring; one of the best and most popular Wiegelias.

Rosea Variegata. A variety of the preceding, with variegated foliage; of dwarfish habit and an exquisite bloomer. 50 cents each.

PRIVET. A class of shrubs or low trees, fine both in foliage and flower.

California. Rich glossy foliage; white flowers.

Japan. Fine shrub or low tree; foliage broad; panicles of white flowers, followed by purple berries. 50 cents each.

Common European. The leaves are dark green and smaller than those of the preceding varieties. Beautiful as a single shrub or for hedging.

HARDY CLIMBING PLANTS.

Except where noted, 25 cents each.

AKEBIA Quinata. A Japanese climber with reddish brown flowers. 35 cents each.

ARISTOLOCHIA Sipho. (Dutchman's Pipe.) Strong growing; has interesting, roundish light green leaves, eight or ten inches in diameter, and curious pipe-shaped, yellowish brown flowers. 75 cents each.

AMPELOPSIS Veitchii. A small leaved variety, which will cling closely to the smoothest walls or boards; foliage very rich and fine; a desirable new variety. 50 cents each.

AMPELOPSIS Quinquefolia. (American Ivy or Virginia Creeper.) A rapid climber, with large five-lobed leaves, which change to the brightest scarlet or crimson in autumn.

BIGNONIA Grandiflora. (Trumpet Creeper.) A splendid climber; vigorous and hardy, with clusters of large scarlet, trumpet-shaped flowers all through the summer and autumn. During the severe drouth in Western Texas in 1886, we saw this growing and blooming beautifully without water or care, and where nearly all other vegetation was parched up.

HONEYSUCKLES. Who does not know and love the sweet Honeysuckles! No other climber is so popular.

Chinese Evergreen. White, buff, and pink; delightfully fragrant; an old reliable sort.

Golden Veined. (*Aurea Reticulata.*) White and cream color; foliage variegated with yellow veins and blotches. Very pretty at all seasons of the year. 35 cts. each.

Hall's. White and buff; very fragrant and a constant bloomer. One of the best. 35 cents each.

Coral. A well known strong growing variety, with showy scarlet flowers.

Yellow Trumpet. Similar to above, differing only in color. 35 cents each.

IVY, English. The well known evergreen climber. Of slow growth, but hardy.

JASMINUM Nudiflorum. Produces light yellow flowers in early spring. 35 cts. each.

Officinalis. Flowers white, fragrant. 35 cents each.

PERIPLOCA Græca. Grecian Silk Vine; rapid grower, flowers purple.

CLIMBING ROSES. Several varieties, Noisettes, etc. See ROSES, page 10.

WISTARIA, Chinese Purple. A beautiful climber of rapid growth, and producing long pendulus clusters of pale blue flowers. When well established makes an enormous growth; it is very hardy and one of the most superb vines ever introduced. 35 cents each.

Chinese White. Flowers pure white; beautiful and rare. 75 cents each.

Honeysuckle.

Abronia.

Aquilegia.

Amaranthus Tricolor.

Flower Seeds.

Sent free by mail on receipt of price. Full cultural directions on each packet.

CLUBBING RATES ON SEED IN PACKETS.

For a remittance of	$1.00 you may select seeds in packets valued at	$1.25
"	" 2.50 " " " "	3.50
"	" 5.00 " " " "	7.50

ABRONIA.

Trailing verbena-like plants; very fragrant. 5 cents.

ACONITUM NAPELLUS.

Showy hardy perennial, growing well in shaded places. 5 cents.

ACROCLINIUM.

Beautiful everlasting flowers. 5 cents.

Album. Pure white.
Roseum. Bright rose.
All Colors Mixed.

AGERATUM.

Whether grown for summer flowers or for winter blooming, this plant keeps up an almost constant bloom. 5 cents.

Mexicanum Nanum. Dwarf blue.
" **White.**
" **Tom Thumb.**

ALYSSUM.

An old favorite, flowering continuously; pretty and fragrant.

Sweet. Flowers white. 5 cents.
Benthami Compactum. Dwarf white. 10 cents.
Wierzbecki. Perennial; yellow. 5 cts.

AMMOBIUM.

A white everlasting flower. 5 cents.

AMARANTHUS.

Grown entirely for their foliage; very ornamental and striking.

Atropurpureus. Blood red foliage. 5 c.
Tricolor. (Joseph's Coat.) Leaves red, yellow and green. 5 cents.
All Colors Mixed. 5 cents.

ANTIRRHINUM.

Beautiful spikes of bright colored flowers, produced all summer and should be in every garden. We offer only the best varieties.

Tom Thumb. Dwarf growing. Mixed, 5 cents.
Finest Mixed. 5 cents.

AQUILEGIA.

(Columbine.)

Exceedingly showy plants, and among the best for early summer blooming.

Alba Flora Plena. Double white. 10 c.
Mixed Double. All colors. 10 cents.

AURICULA.

Finest Mixed, 15 cents.

Aster, Victoria.

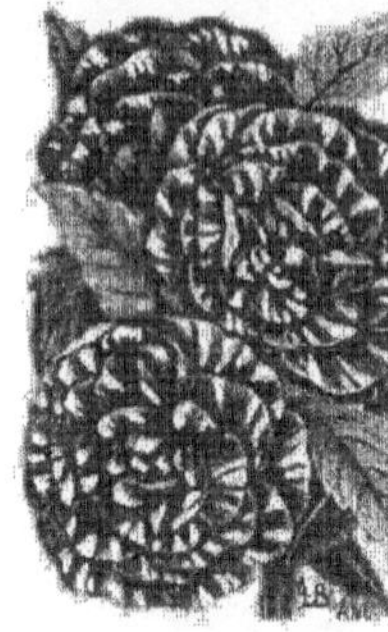

Balsam, Camellia-Flowered.

Calliopsis or Coreopsis.

ASTERS.

No family of plants bears such distinct marks of progress as the Aster, and none are more eagerly sought; one of the most effective of our garden favorites, producing in profusion flowers in which richness and variety of color are combined with the most perfect and beautiful form.

China. Fine mixed. 5 cents.
Chrysanthemum-Flowered. Dwarf mixed. 10 cents.
Comet, New Dwarf. Resembles the Japanese Chrysanthemum, with long curved petals; light pink, bordered white. 25 c.
Dwarf Bouquet. 10 cents.
Pæony-Flowered Globe. Large and double. 10 cents.
Reid's German Quilled. 5 cents.
Truffaut's Pæony-Flowered. Mixed colors. 10 cents.
Victoria. Finest mixed. 10 cents.

BALSAMS.

(Lady's Slipper.)

The Balsam has been so much improved as to be scarcely recognized.

Camellia-Flowered, White. 10 cents.
" " **Pink.** 5 cents.
" " **Purple.** 5 cents.
" " **Scarlet.** 5 cents.
" " **Finest Mixed.** 5 c.
" " **Extra Choice Mixed.** 10 cts.

BELLIS.

(Double Daisy.)

Finest Mixed. Beautiful. 10 cents.

BROWALLIA.

Handsome and profuse blooming plants covered with pretty blue and white flowers throughout the summer. 5 cents.

CALENDULA.

Hardy annual; free flowering and attractive.

Pluvialis. (Cape Marigold.) White. 5 c.
Pongei. (Pot Marigold.) Double white. 5 cents.

CALLIOPSIS or COREOPSIS.

Beautiful and showy plants, flowering freely.

Tall, Mixed. 5 cents.
Tom Thumb, Mixed. 10 cents.

CANARY BIRD FLOWER.

Desirable rapid climber; yellow. 5 cts.

CAMPANULA.

A favorite class of plants.

Lorei. Mixed, white and lilac. 5 cents.
Media. (Canterbury Bells.) Mixed. 5 c.

CANDYTUFT.

Popular and sweet; easy of culture.

New White. 5 cents.
Rocket. Pure white. 5 cents.
Fine Mixed. 5 cents.
Tom Thumb, Mixed. 5 cents.

CARNATIONS.

Double Mixed. 10 cents.

CELOSIA.

(Cockscomb.)

One of the most satisfactory and showy plants for garden decoration.

Cristata. Dwarf crimson. 5 cents.
Dwarf Mixed. 5 cents.
Glasgow Prize. Dark crimson; large. 10 cents.

CENTAUREA. (Dusty Miller.)

The plants are among the best silver-leaved varieties in cultivation.

Candidissima. The dwarfest silver-leaved plant used for ribbon border. 10 cents.
Gymnocarpa. A graceful silver foliaged variety; one of the best for bedding purposes. 10 cents.

CHRYSANTHEMUM.

Mixed Annual Varieties. 5 cents.
Frutescens Grandiflorum. The "Marguerite" or "Paris Daisy." 10 cents.

Clarkia.

Convolvulus.

Dianthus.

CINERARIA MARITIMA.

Silver-leaved foliage. 5 cents.

CLARKIA.

Pretty, cheerful looking flowers.

Finest Mixed Varieties. 5 cents.

CLEMATIS.

Well known climber.

Flammula. White; very pretty. 5 cts.

CONVOLVULUS.

Morning Glory. Mixed. 5 cents.

COSMOS.

A noble race of plants which attain a height of nearly five feet, and which, in the fall months, are literally covered with flowers which closely resemble single Dahlias. The flowers are from one to two inches in diameter, and range through all shades of rose, purple, flesh color and pure white. 10 cents.

CUPHEA. (Cigar Plant.)

Rœzlii Grandiflora. 10 cents.

CYCLAMEN.

Valuable winter-blooming greenhouse plant. 15 cents.

CYPERUS, VARIEGATED.

(Umbrella Plant.)

A strikingly handsome foliage plant, easily raised from seed. 10 cents.

CYPRESS VINE.

Mixed. 5 cents.

DELPHINIUM.

(Perennial Larkspur.)

Nudicaule. Scarlet. 10 cents.
Formosum. Rich blue and white. 5 cents.

DAHLIA.

Double Choice Mixed. 10 cents.
New Single. 20 cents.

DIANTHUS.

This magnificent tribe is one of the most satisfactory that can be raised from seed; flowers of the most brilliant color.

China or Indian Pink. 5 cents.
Alba Flora Plena. Double white. 5 cts.
Heddewigii fl. pl. Mixed colors. 10 cts.
Barbatus. (Sweet William.) 10 cents.
Plumarius. (Pheasant's Eye.) 10 cents.

DIGITALIS. (Foxglove.)

Handsome ornamental plant.

Mixed Colors. 5 cents.

ERYSIMUM.

Showy handsome annual.

Arkansanum. (Western Wallflower.) Yellow. 5 cents.

ESCHSCHOLTZIA.

(California Poppy.)

Fine Mixed. All colors. 5 cents.

EUCALYPTUS. (Blue Gum Tree.)

Globulus. (Fever and Ague Plant.) Not hardy. 10 cents.

EUPATORIUM.

Fraserii. 5 cents.

GERANIUM.

Fine Mixed Varieties. 20 cents.
Apple Scented. (True.) Very fragrant. 15 cents.

GLADIOLUS.

French Hybridized Seed. 10 cents.

GLOBE AMARANTH.

(Bachelor's Buttons.)

Remarkably handsome everlastings.

White, Purple and Mixed. 5 cents.

Gourds.

Helianthus, "Oscar Wilde."

Lantana.

GODETIA.

Well worthy extended cultivation; their delicates tints of purple and pink have made them famous.

Lady Albemarle. Flowers large, measuring over three inches across. 5 cents.
The Bride. White, with crimson eye. 5 cents.
Fine Mixed. 5 cents.

GRASSES. (Ornamental.)

Briza Maxima. Quaking grass. 5 cents.
Bromus Brizæformus. A fine grass. 5 c.
Stipa Pennata. Feather grass. 10 cts.

GOURDS. (Ornamental.)

Dipper. 5 cents.
Mixed. 5 cents.

HELIANTHUS. (Sunflower.)

Californicus fl. pl. Large and double. 5 cents.
Oscar Wilde. Flowers small, with small jet black center, surrounded by a row ol bright golden petals. 5 cents.
Peruvianus. Double yellow flowers, striped black. 5 cents.
Silver Leaf. Downy, silvery foliage; flowers yellow and black. 5 cents.
Choice Mixed. 5 cents.

HELICHRYSUM.

An everlasting; used mostly for winter bouquets. 5 cents.

HIBISCUS.

Californicus. Pure white; constant bloomer. 5 cents.
Africanus. Rich cream, with brown center. 5 cents.
Moschentos Roseus. Large, rose color. 5 cents.

HOLLYHOCK.

The old garden favorite, much improved.

Choice Mixed. 10 cents.
Double White. 10 cents.

HONESTY. (Lunaria.)

Purple. 5 cents.

HUMULUS JAPONICUS.

A new hop from Japan. A rapid grower and very ornamental vine. 25 cents.

ICE PLANT.

Trailing plant, suitable for vases or baskets. 5 cents.

IMPATIENS SULTANI.

Good for the greenhouse or open ground. Plant compact and almost a perpetual bloomer. Flowers brilliant rose-scarlet and borne so profusely on the plant, that a well grown specimen appears to be a perfect ball of flowers. 15 cents.

IPOMEA.

Bona Nox. (Moonflower.) 10 cents.
Coccinea. Bright scarlet. 5 cents.

LANTANA.

Very showy; blooms all summer 5 cts.

LARKSPUR.

Well known and hardy.

Dwarf Rocket. Finest mixed double. 5 cents.
Tall Rocket. Finest mixed double. 5 cts.

LATHYRUS. (Everlasting Pea.)

Ornamental, free flowering rapid climber; flowers pretty. 5 cents.

LOBELIA.

Dwarf plants admirably adapted for the front lines of ribbon borders and for vases and hanging baskets.

Speciosa. One of the most effective for bedding. 5 cents.
Crystal Palace, Deep blue. 5 cents.
Finest Mixed. 5 cents.
Cardinalis. Long spikes, vermilion-scarlet. 5 cents.

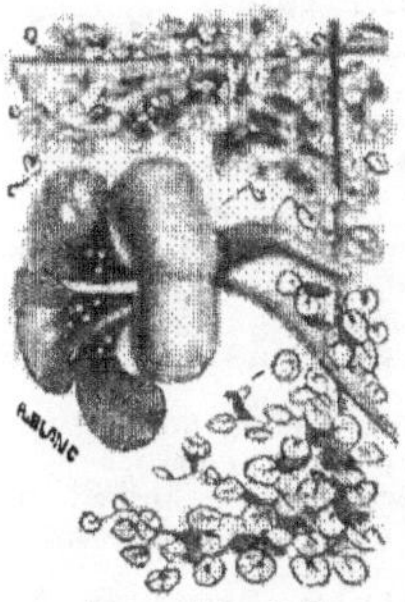

Nasturtium, Tall.

Pansy, Good Mixed.

Phlox Drummondii.

LUPINS.

Desirable plants for every garden.

Mixed. 5 cents.

MARIGOLD.

African Quilled. Orange and yellow mixed. 5 cents.
New African El Dorado. New and beautiful, with great range of colors. 5 cents.
Dwarf French. Mixed. 5 cents.
New Dwarf French. Gold-striped ; rich maroon, striped with golden yellow. 5 c.

MARVEL OF PERU.
(Four O'Clocks.)

Mixed. 5 cents.

MAURANDIA.

Graceful climber ; valuable for both flower and foliage, and useful for hanging baskets.

Barclayana. Rich violet. 10 cents.
Finest Mixed. 10 cents.

MIGNONETTE.

An old favorite ; blooms profusely all summer.

Large Flowered. 5 cents.
Pyramidal. 5 cents.
Machet. Plant dwarf ; flowers red and deliciously scented. 10 cents.

MIMULUS.
(Monkey Flower.)

Cardinalis. Scarlet. 5 cents.
Moschatus. (Musk Plant.) Yellow. 5 c.

MYOSOTIS.
(Forget-Me-Not.)

Palustris. Blue. 10 cents.
Alba. White. 5 cents.
Azorica Cœlestina. Turquoise blue. 5 c.

NEMOPHILA.

Of very compact habit, and if planted three or four inches apart will present a dense mass of flowers.

Insignis. Bright blue, white center. 5 c.

NASTURTIUM.

Nasturtiums will always be general favorites, for the reason that they stand any amount of heat and drought, growing vigorously and flowering freely, especially in a poor, rocky soil.

Tom Thumb, Scarlet. 5 cents.
" " **Spotted.** 5 cents.
" " **Ruby King.** 10 cents.
" " **Crystal Palace Gem.** Sulphur yellow, maroon spots. 5 cents.
" " **King of Tom Thumbs.** Crimson scarlet. 5 cts.
" " **Mixed.** 5 cents.
Tall Varieties Mixed. 5 cents.

OXALIS.

Beautiful plants, suitable for rock-work and rustic baskets. 10 cents.

PANSY.

Finest Mixed. 10 cents.

PENTSTEMON.

Fine Mixed. 10 cents.

PETUNIA.

For out-door decoration in summer, scarcely any plant equals this.

Striped and Blotched Mixed. 10 cts.
Countess of Ellesmere. Pink. 5 cts.
Double Mixed. 10 cents.

PHLOX DRUMMONDII.

One of the finest of annual plants, and stands almost unrivalled for profusion and duration of bloom and richness of color.

Alba. Pure white. 5 cents.
Black Warrior. Dark purple. 5 cents.
Coccinea. Deep scarlet. 5 cents.
Leopoldii. Bright rose, white eye. 5 cts.
Princess Royal. Light purple, white streaks. 5 cents.
Queen Victoria. Violet, white eye. 5 cents.
Rosea. Pure rose. 5 cents.
Finest Mixed. 5 cents.

PERILLA.

Black purple-colored foliage plant. 5 cts.

POPPY.

Showy, desirable and easily grown.

Double Carnation-Flowered, Mixed. 5 cents.

Pæony-Flowered, Mixed. 5 cents.

Choicest Mixed. 5 cents.

PORTULACA.

Beautiful and popular hardy annuals, of the easiest culture, luxuriating in an exposed sunny situation, and producing through the summer their flowers of every hue in the greatest profusion.

Single Varieties, All Colors Mixed. 5 c.

Double Varieties, All Colors Mixed. 10 cents.

PYRETHRUM.

Golden Feather. Bright golden foliage; one of the best bedding plants. 5 cents.

Golden Gem. Foliage brighter than the *Golden Feather;* flowers double; white. 10 cents.

RHODANTHE ATROSANGUINEA.

(Everlasting Flower.)

New Double. The plant grows a foot high, and produces in abundance, perfectly rosy carmine double flowers. 15 cents.

ROCKET.

Mixed. Pretty flowers. 5 cents.

SALVIA SPLENDENS.

(Flowering Sage.)

Bright red; fine in the fall. 10 cents.

SEDUM CŒRULEUM.

A charming little trailing plant useful for rock-work, etc.; flowers bright blue. 5 c.

STATICE.

Easily cultivated; free flowering.

Spicata. Rosy pink. 10 cents.

STOCKS.

(Gilliflower.)

German Ten Weeks—
- **Mixed.** 5 cents.
- **Large-Flowered, Mixed.** 10 cents.
- **Wallflower-Leaved.** Pure white. 10c.

Brompton or Winter, Mixed. 10 cents.

Large-Flowered Dwarf German Ten Weeks. In six separate colors. 50 cts.

Intermediate. In six separate colors. 50 c.

SWEET PEAS.

Grown on hedges or sticks, or for trellis work, they are almost unsurpassed. Their great variety of color and fragrance make them desirable in every garden.

Adonis. A lovely shade of rose.

Butterfly. Pure white ground, delicately laced with lavender blue. 5 cents.

Black Purple. Black. 5 cents.

Crown Princess of Prussia. Bright blush. 5 cents.

Miss Ethel. Delicate pink, flushed with crimson, blush wings. 5 cents.

Queen of the Isles. Scarlet, flushed and mottled, with white wings, flaked and margined rosy purple. 5 cents.

Scarlet Invincible. Remarkably fragrant, with bright scarlet crimson flowers. 5 cents.

White. 5 cents.

Mixed. All colors. 5 cents.

SWEET WILLIAM.

Fine Mixed. 5 cents.

TROPÆOLUM.

Elegant and beautiful climber, flowering freely in the open ground.

Brilliant Mixed Varieties. 5 cents.

VERBENA.

Saved from the best named varieties.

Scarlet, White, Striped and Mixed. 10 cents.

VINCA.

(Madagascar Periwinkle.)

Ornamental free flowering plants, blooming continuously through the hottest and dryest summer.

Rosea. Rose, with dark eye. 10 cents.

Alba. White, with crimson eye. 10 cents.

Alba Pura. Pure white. 10 cents.

VIOLA ODORATA.

Sweet violet, blue. 10 cents.

WALLFLOWER.

Double Mixed. 10 cents.

Single Mixed. 5 cents.

XERANTHEMUM.

Double White Everlasting. 5 cents.

ZINNIA.

Double Mixed. Choice. 5 cents.

Double Striped. 5 cents.

Vegetab e Seeds.

We base prices in this list without postage paid. The purchaser will kindly add postage when buying seed in bulk to be sent by mail, at the rate of eight cents for every pound of seed. For postage on Peas and Beans, please add fifteen cents per quart, and on Corn ten cents per quart. Ounce packages or packets will be mailed free at quoted prices.

ASPARAGUS.

	Pkt.	oz.	¼ lb.	lb.
Conover's Colossal	$0 05	$0 10	$0 20	$0 55

BEET.

Bastian Early Turnip	5	10	20	75
Dewing's Early Blood	5	10	20	75
Eclipse, Best Early	5	10	30	1 00
Long Blood	5	10	20	75
Mangel Wurzel, Red	5	10	20	50

BEANS.

		qt.	pk.	bu.
Cleveland's Improved Red Valentine	5	$0 30	$2 00	$6 50
Best of All	5	35	2 00	6 50
Golden Wax	5	30	2 00	6 50
Pole, Southern Prolific	5	30	2 00	6 00

CORN.

Cory, the earliest	5	30	1 50	5 00
Early Minnesota	5	20	1 25	4 00
Stowell's Evergreen	5	20	1 25	4 00
Black Mexican	5	20	1 25	4 00
Mammoth Late	5	20	1 25	4 00

CABBAGE.

		oz.	¼ lb.	lb.
Early Drumhead	5	$0 15	$0 50	$1 50
" Winnigstadt	5	15	50	1 50
" York, Large	5	15	50	1 50
Extra Early Etampes	5	15	50	1 50
Fottler's Brunswick	5	20	60	2 00
Jersey Wakefield	5	20	50	1 75
Large Late Drumhead	5	25	75	2 50
Premium Flat Dutch	5	15	50	1 75

CARROT.

Half Long Red	5	10	30	1 00
Long Orange	5	10	25	75

CORN SALAD or FETTICUS.

Large Seed	5	10	25	1 00

CELERY.

Giant White Solid	5	20	75	2 50

CUCUMBER.

Early Frame	5	10	30	1 00
Green Prolific	5	10	30	1 00
Long Green Turkey	5	10	30	1 00
White Spine Improved	5	10	30	1 00

EGG PLANT.

New York Improved	5	40	1 25	4 00

LETTUCE.

	Pkt.	oz.	¼ lb.	lb.
Black Seeded Simpson	$0 05	$0 20	$0 50	$1 50
Boston Curled	5	20	50	1 50
Early Curled Simpson	5	20	50	1 50
Hanson	5	20	50	1 50

MELON, MUSK.

Casaba	5	10	30	1 00
Jenny Lind	5	10	30	1 00
Pine Apple	5	10	30	1 00

MELON, WATER.

Georgia Rattlesnake or Gypsy	5	10	30	1 00
Kolb's Gem	5	10	30	1 00

MUSTARD.

Black	5	10	15	30
Southern Giant Curled	5	10	30	1 00

OKRA or GUMBO.

Dwarf White	5	10	25	75
Tall White	5	10	25	75

ONION.

Extra Early Red	5	20	60	2 00
White Portugal or Silver Skin	5	30	75	3 00
Yellow Globe Danvers	5	20	60	2 00

PARSLEY.

Champion Moss Curled	5	10	30	1 00
Double Curled	5	10	30	1 00
Fern-Leaf	5	10	20	1 00

PEAS.

		qt.	pk.	bu.
Cleveland's Alaska	5	$0 30	$2 00	$7 00
" First and Best	5	25	1 75	6 00
Carter's Premium Gem	5	25	1 50	5 00
Laxton's Alpha	5	25	1 50	5 00
Carter's Strategem	5	40	2 50	8 00
Cleveland's L. I. Mammoth	5	40	2 50	8 00
Everbearing (Bliss')	5	35	2 00	7 00

PEPPER.

		oz.	¼ lb.	lb.
Chili, Small Red	5	$0 30	$1 00	
Large Bell or Bull Nose	5	30	1 00	
Sweet Mountain	5	30	1 00	

PUMPKIN.

Cushaw	5	10	20	$0 60

RADISH.

Early Long Scarlet Short Top	5	10	25	75
" Scarlet Turnip, White Tipped	5	10	25	75
French Breakfast	5	10	25	75
New White Strasburgh	5	10	25	75
Wood's Early Frame	5	10	25	75

SPINACH.

Bloomsdale	5	10	15	40
Thick-Leaved Round	5	10	15	40

SALSIFY or VEGETABLE OYSTER.

Sandwich Island Mammoth	10	40	1 35	

SQUASH.

Hubbard	5	10	25	75
White Bush Scolloped	5	10	25	75

TOMATO.

	Pkt.	oz.	¼ lb.	lb.
Acme	$0 05	$0 20	$0 70	$2 50
Livingston's Beauty	5	20	70	2 50
" Perfection	5	20	70	2 50
Optimus	5	30	1 00	3 50

LAWN GRASS SEED.

	qt.	pk.	bu.
Emerald	20	1 25	4 00
Central Park	15	1 00	3 00

GRASS AND CLOVER SEEDS.

Price of grass seeds are variable and will be given on application.

Johnson Grass.
Kentucky Blue Grass.
Timothy or Herd Grass.
German or Golden Millet.
Hungarian Grass or Millet.
Alfalfa or Lucerne.
Red Clover.
White Clover.

Cleveland's Improved Red Valentine Bean.

INDEX.

GREENHOUSE DEPARTMENT.

FRUIT DEPARTMENT.

ORNAMENTAL DEPARTMENT.

SEED DEPARTMENT.

www.ingramcontent.com/pod-product-compliance
Lightning Source LLC
Chambersburg PA
CBHW022006190326
41519CB00010B/1411
9783741136276